Revisiting Sustainable Tourism in the Philippines: Towards a Better Normal

NEW PERSPECTIVES IN TOURISM AND HOSPITALITY MANAGEMENT

New Perspectives in Tourism and Hospitality Management positions organizational change and behavior in the tourism and hospitality sector at its heart, providing interdisciplinary research in this field and joining the areas of organizational studies and tourism/hospitality management.

After years of disruption, the tourism and hospitality sector is an increasingly important area of study, encompassing topics such as sustainability, the environment, climate change, terrorism, cultural studies, and more. New trends, challenges, and developments, as well as the application of new ideas that are likely to affect the tourism and hospitality industry are analyzed with a particular focus on organizational behavior, responsible management and governance, sustainability, and ecotourism.

New Perspectives in Tourism and Hospitality Management provides alternative philosophies for organizational change and development in hospitality. It encourages the exploration of philosophies including critical theory and is interested in qualitative and quantitative analyses of change management and change practices in touristic destinations and organizations.

Previous Volumes

Sport and Tourism: Strategies to Develop Tourist Destinations
Edited by: Marco Valeri

Tourism Innovation in the Digital Era: Big Data, AI and Technological Transformation
Edited by Marco Valeri

Forthcoming Volumes

Destination Conscience: Seeking Meaning and Purpose in the Travel Experience
Edited by: Harald Pechlaner, Natalie Olbrich, and Giulia Isetti editing on behalf of EURAC

The Regenerative Tourism Industry: The Relationship Between Gender Equality, Tourism
Edited by: Geetanjali Sageena and Suneel Kumar

Competitiveness of Tourism SMEs in Post Pandemic Times: A Multidisciplinary Analysis
Editor by: Antonio Emmanuel Pérez Brito and Martha Isabel Bojórquez Zapata

Revisiting Sustainable Tourism in the Philippines: Towards a Better Normal

EDITED BY

JOHN PAOLO R. RIVERA

Philippine Institute for Development Studies, Philippines; Asian Institute of Management - Dr. Andrew L. Tan Center for Tourism, Philippines

EYLLA LAIRE M. GUTIERREZ

Asian Institute of Management - Dr. Andrew L. Tan Center for Tourism, Philippines

AND

FERNANDO MARTIN Y. ROXAS

National Power Corporation, Philippines; Asian Institute of Management - Dr. Andrew L. Tan Center for Tourism, Philippines

emerald PUBLISHING

United Kingdom – North America – Japan – India – Malaysia – China

Emerald Publishing Limited
Emerald Publishing, Floor 5, Northspring, 21-23 Wellington Street, Leeds LS1 4DL

First edition 2024

British Library Cataloguing in Publication Data
A catalogue record for this book is available from the British Library

ISBN: 978-1-83753-679-5 (Print)
ISBN: 978-1-83753-678-8 (Online)
ISBN: 978-1-83753-680-1 (Epub)

Printed and bound by CPI Group (UK) Ltd, Croydon, CR0 4YY

Contents

List of Figures and Tables

Figures

Tables

About the Editors

John Paolo R. Rivera is an Educator, Researcher, and Consultant in the areas of economics, management, and tourism. He is the President and Chief Economist of Oikonomia Advisory & Research, Inc. and is a Senior Research Fellow at the Philippine Institute for Development Studies. He was OIC Executive Director at the AIM-Dr. Andrew L. Tan Center for Tourism.

Eylla Laire M. Gutierrez is a Program and Research Manager at the Asian Institute of Management (AIM) Dr Andrew L. Tan Center for Tourism. Her research interests include sustainability, community development, gender studies, and tourism management.

Fernando Martin Y. Roxas is the President of the National Power Corporation and a Professor at the Asian Institute of Management, where he teaches operations and services management, supply chains, systems thinking, and sustainable tourism.

About the Contributors

Luisito C. Abueg is an Assistant Professor at the University of the Philippines Los Baños (UPLB). His research interests are mathematical economics, economic statistics, economic history, and gender economics.

Iris L. Acejo is a Research Associate with Seafarers International Research Center in the School of Social Sciences at Cardiff University. Her research interests are welfare and vulnerability of seafarers.

Ramon Benedicto A. Alampay is an Associate Professor at the UP Asian Institute of Tourism (AIT). His research interests are local and regional economic development, regenerative approaches to placemaking, and tourism strategy.

Avi Ben P. Andalecio is an Academic Staff and a Researcher at the University of Santo Tomas (UST) College of Tourism and Hospitality Management (CTHM). His research interests are tourism and hospitality, public policy, and administration.

Warner M. Andrada is the Department of Tourism (DOT) OIC-Assistant Secretary for Tourism Development. His research interests are tourism policies and plans, analysis of tourism data, and measuring tourism impact on the economy.

Lilibeth C. Aragon is the Dean of the Lyceum of the Philippines University (LPU) Manila's College of International Tourism and Hospitality Management.

Jonna C. Baquillas is an Associate Professor at the Department of Marketing and Advertising of De La Salle University (DLSU). Her research interests are sustainable consumption and production, circular economy, sustainable tourism, climate resilience, and just energy transition.

Marie Jel D. Bautista was an Assistant Professor at the Department of Hospitality and Tourism Management of Central Luzon State University (CLSU), Nueva Ecija.

Ma. Janda Ira Felina M. Benedictos is an Instructor at UPLB handling undergraduate courses on elementary economics, macroeconomics, microeconomics, and human resource economics.

Jocelyn Y. Camalig is the Chairperson of the Hotel and Restaurant Administration at LPU Manila. She is a Certified Hospitality Educator, Certified Guest

Service Professional, ServSafe Practitioner, and National Master Trainer/ Assessor for the Food and Beverage Service Division.

Celyrah B. Castillo is a Professor and the Dean of the College of Home Science and Industry at CLSU. She worked in various research and development projects on internationalization of Higher Education Institution through education tourism.

Daryl Ace V. Cornell is a Professor at the Polytechnic University of the Philippines (PUP). His research interests are sustainable operations in tourism and hospitality and leisure and recreation management.

Ethelbert P. Dapiton is a Visiting Professor at Swinburne University of Technology, Victoria, Australia. His research interests are education, tourism, and hospitality management.

Edieser D. L. Dela Santa is a Professor at the UP AIT. His research interests are politics of tourism policy formulation and implementation, representation of culture and heritage, tourism and hospitality curriculum development, and tourism governance in protected areas.

April Joy E. Dopeño is an Academic Staff at the Department of International Studies, Institute of Arts and Sciences of FEU Manila.

Omme Atiyah B. Gonting is a Tourism Operations Officer at the Provincial Tourism Office of La Union, Philippines and a Master's student at the UP Asian Institute of Tourism. Her research interests are sustainable tourism value chains, destination marketing, and cultural tourism.

Gezzez Giezi G. Granado is the Dean of UST CTHM and a Professor of Law at UST CTHM and UST Faculty of Civil Law. His research interests are sustainable tourism, legal policy issues in tourism and hospitality, and tourism and hospitality education.

Marie Danielle V. Guillen is an Associate Professor at UP AIT and is an International Development Professional and a Consultant in the field of policy and planning sciences, specifically transport, tourism, gender, and climate change.

Viory Yvonne T. Janeo is the Vice Dean and Faculty at University of Asia & the Pacific (UAP) School of Economics.

Jovito Jose P. Katigbak is a Lecturer at the Department of Political Science and Development Studies of DLSU Manila. He is a Senior International Finance Specialist at the Bangko Sentral ng Pilipinas. His research interests are women and development, sustainable development, e-commerce, governance and public policy, and international relations.

Liwliwa B. Lagman is an Associate Professor of Tourism and Hospitality at University of Mindanao. Her industry experience includes managing daily operation of front office and housekeeping departments as Hotel Duty Manager.

Jame Monren T. Mercado is an Academic Staff and a Researcher at UST CTHM. His research interests are cultural heritage tourism management and sustainable development; and intangible cultural heritage.

Winston Conrad B. Padojinog is the President of UAP. His research interests are industrial dynamics and policy.

Ivy Charisse T. Pagulayan is the Chairperson at LPU Manila's College of International Tourism and Hospitality Management. Her industry experience includes working as Front Desk Officer, Barista, and Restaurant Crew at Portofino Bay Hotel, Orlando, Florida.

Diana Rueda is an Instructor at UAP. She teaches International Business Economics, Intermediate Microeconomics, and Macroeconomics.

Claire Therese B. Villafuerte is a Teaching Associate at the UPLB College of Economics and Management. Her research interests are economic development, human resource economics, and gender economics.

Ray Justin A. Villanueva has an undergraduate degree in Economics with an informal minor in Tourism Management from UP Diliman. His areas of interest include tourism marketing and value chains, redesigning of local tourism marketing campaigns, and assessment of community tourism policies.

Foreword

Fernando Martin Y. Roxas

We are pleased to introduce "Revisiting Sustainable Tourism in the Philippines: Towards a Better Normal." In a world marked by dynamic shifts and unprecedented challenges, this book delves into the transformative journey of sustainable tourism in the Asian context, examining its evolving landscape against the backdrop of technological advances, the COVID-19 experience, and the increasing sensitivity to sustainability.

Technology is a powerful enabler for sustainable tourism. From digital platforms that connect travelers with eco-friendly accommodations to advanced data analytics aiding in destination management that reduce our carbon footprint, technology is empowering both businesses and tourists to make responsible choices.

The shockwaves of the pandemic have reverberated globally, prompting the opportunity to prioritize resilience and adaptability in the face of unforeseen challenges. We explore how the lessons learned during the global crisis can be harnessed to create a tourism sector that is more adaptable, inclusive, and considerate of local communities.

Our society's heightened awareness to sustainability is another beacon guiding the evolution of tourism. On one side, communities, governments, and businesses are recognizing the necessity to balance economic benefits with environmental and social responsibility. On the other side, conscious travelers are increasingly seeking experiences that align with their values, driving a demand for destinations that support community engagement, cultural preservation, and the protection of our planet.

We tackled postpandemic issues hindering sustainability pursuits in the better normal. Tourism management and development have become more profound with the growing significance of tourism as an economic pillar, particularly in Asia – a region teeming with ridge-to-reef experiences. Asia is home to major tourism and investment source markets such as the People's Republic of China (PRC), Republic of Korea (ROK), Japan, India, and the member economies of the Association of Southeast Asian Nations (ASEAN). Economic opportunities from tourism brought about by hyper-globalization are driving changes in the strategies of national tourism organizations (NTOs), policymakers, destination managers, private enterprises, local communities, and tourists. The challenges of meeting these opportunities have become more pronounced given consequences of unsustainable tourism practices. Thus, we present challenges, opportunities, and approaches of pursuing sustainable tourism in a postpandemic situation within the framework, principles, and practices of tourism management and development as applied to the diverse yet fragile nature-based tourism sector of the Philippines. We showcase an anthology of research focusing on the Philippine experience, which other economies can learn from.

In 2005, the Philippine Institute of Development Studies (PIDS) – Philippine APEC Study Center Network (PASCN) published the book *Challenges of Sustainable Tourism in the Philippines*.[1,2] A handful books on tourism economics and economics of sustainable tourism also presented case studies from developed and less developed tourism destinations. They focused on market demand estimations, host communities, and environmental impact analysis. However, in 2023, the Asian Institute of Management – Dr Andrew L. Tan Center for Tourism (ALT-CFT), through De La Salle University Publishing House (DLSUPH), released the book *People, Planet, Profit: Principles and Practices of Sustainable Tourism* highlighting more Philippine and Asian cases.[3,4] As a follow-up, we continue the trajectory of underscoring less developed tourism destinations and less discussed issues, which are value-adding for policymakers, private sector, local community, civil society, academe, and the public.

We structured this book to guide you through essential chapters, including a fresh and in-depth preliminary exploration of sustainable tourism, a paradigm shift toward value chain integration, a call to think small for effective management, the customization of services to meet changing demands, and a compelling conclusion advocating for quality over quantity. It adopts a unique management and practical perspective. By merging theoretical frameworks with real-world experiences, our contributors offer tangible insights that can be applied by professionals, policymakers, and stakeholders alike. The book analyzes the current state of sustainable tourism and contributes to the discourse on how it can be realized.

We thank the contributors who shared their insights, experiences, and expertise. We hope that this encourages you to reflect on the profound implications of the main themes, foster constructive dialogues, and spark collective action toward a more sustainable and resilient future for tourism.

[1]https://www.pids.gov.ph/
[2]https://pascn.pids.gov.ph/
[3]https://aim.edu/research-centers/dr-andrew-l-tan-center-tourism
[4]https://www.dlsu.edu.ph/research/publishing-house/

Preface

John Paolo R. Rivera

This book tackles the postpandemic theoretical, conceptual, empirical, and systemic issues faced by tourism stakeholders in their pursuit of sustainability in the new and better normal. More than ever, the role of tourism management and development has become more profound with the growing significance of tourism as an economic pillar, particularly in Asia – a region teeming with ridge-to-reef experiences anyone can partake. Moreover, Asia is home to major tourism and investment source markets such as the People's Republic of China (PRC), Republic of Korea (ROK), Japan, India, and the member economies of the Association of Southeast Asian Nations (ASEAN). The economic opportunities from travel and tourism brought about by hyper-globalization are driving changes in the behavior and strategies of national tourism organizations (NTOs), policymakers, destination managers, private business organizations from micro to large enterprises, local communities, and tourists themselves. Likewise, the challenges of meeting these economic opportunities have become more pronounced given the negative repercussions of unsustainable tourism practices on the people and the environment. Thus, this book explicates not only the challenges and the opportunities of pursuing sustainable tourism in a postpandemic situation but also the approaches by which sustainable tourism in Asia can be achieved within the framework, principles, and practices of tourism management and development as applied to the diverse yet fragile nature-based tourism sector of the Philippines. Through this book, we present a collection of peer-reviewed research works focusing on the experiences of the Philippines, which other economies heavy on tourism can learn from.

The key objective in preparing this book is for it to serve as key evidence-based resource material for achieving a broader and deeper understanding of the current challenges and opportunities of sustainable tourism. This is anchored on the lessons learned from the COVID-19 pandemic and how the tourism industry can move forward sustainably and resiliently. It compiles and structures in one anthology pertinent enquiries, issues, solutions, and lessons encountered by stakeholders in driving sustainable tourism. Equally important, this book also reinforces the research culture in the tourism industry. While research can be basic, applied, or developmental, it can generate patents, innovations, and new methodologies that can contribute to the development of the industry in the long run.

While sustainability has become more than a buzzword, this book bridges a significant gap in the literature of sustainability particularly in tourism. There is inadequacy of major scholarly references synthesizing the application of economic and management principles and techniques in a postpandemic sustainable tourism paradigm: (1) a value chain approach to sustainable tourism; (2) "thinking small" as a new pillar of sustainable tourism; and (3) customization to recalibrate tourism products and services. These facets have not yet been covered in the form of a book. Hence, this book offers value-added through exploring the applications of economics and management as well as providing relevant case studies on sustainable tourism in the Philippines that other countries can benchmark with.

In 2005, the Philippine Institute of Development Studies (PIDS) and the Philippine APEC Study Center Network (PASCN) published a book on *Challenges of Sustainable Tourism in the Philippines*.[5,6] In the international market, there are quite a handful of books on tourism economics and economics of sustainable tourism that present case studies from developed tourism destinations and a few from less developed destinations. Chapter topics focus largely on market demand estimations, host communities, and environmental impact analysis. However, in 2023, the Asian Institute of Management – Dr Andrew L. Tan Center for Tourism (ALT-CFT), through De La Salle University Publishing House (DLSUPH), released a book on *People, Planet, Profit: Principles and Practices of Sustainable Tourism* that sought to highlight more of Philippine and Asian case studies.[7,8] As a follow-up, this book continues the trajectory of underscoring less developed tourism destinations and less discussed issues, which can be a useful guide for policymakers, private sector, local community, civil society, academe, and the public.

---

[5]https://www.pids.gov.ph/

[6]https://pascn.pids.gov.ph/

[7]https://aim.edu/research-centers/dr-andrew-l-tan-center-tourism

[8]https://www.dlsu.edu.ph/research/publishing-house/

I

A Value Chain Approach to Sustainable Tourism

Chapter 1

Shifting Paradigms: An Empirical Analysis of the Tourism Value Chains in the Better Normal

Ray Justin A. Villanueva

University of the Philippines Diliman – School of Economics, Philippines

Abstract

The COVID-19 pandemic resulted in significant declines in international tourist arrivals and receipts. It has also influenced destination preference, tourist demographic, travel motivation, and behavior. Recognizing health and safety as the new considerations in pursuing tourism activities in the better normal, this necessitated a reassessment of the current tourism industry by directing the attention from the usual destination-centric perspective to a value chain perspective. This chapter proposes a new paradigm for the better normal value chain by deconstructing the concepts of travel, tourism, and travel sectors and revisiting the concept of the tourism value chain (TVC) by mapping out the chain and its functional levels and integrating travel, tourism, and hospitality sectors in one value chain. Policymaking approaches such as reorganizing the value chain, empowering stakeholder involvement through coopetition, and resilience building in the face of possible adversities in the future should be adapted to achieve this suggested paradigm's goals. This analysis provides stakeholders with a broader understanding of the needed interventions in future-proofing the industry backed by industry trends in the better normal while fostering collaboration and offering flexibility to cope better in other possible shocks in the future.

Keywords: ASEAN tourism; value chain; coopetition; better normal; functional levels

Revisiting Sustainable Tourism in the Philippines, 3–29

doi:10.1108/978-1-83753-678-820241001

Impact of COVID-19 on ASEAN Tourism

The COVID-19 pandemic highlighted the various weak links in the various global value chains *affecting economies and industries, as well as the unprecedented changes in the demand and* supply bases of these industries (Tasnim et al., 2022). While many industries have been drastically affected by the pandemic, the global TVC has been one of the deadliest hit due to the repercussive consequences of the border restrictions, paralyzing labor markets, hospitality and tourism segments, leading to changes in traveler behavior and tourism preferences, evidenced by documented literature of Chakraborty and Kar (2021), Kaushal and Srivastava (2021), and Lew et al. (2020), leaving a profound impact on the economies of many economies, as tourism is a significant source of employment and revenue especially in the Association of Southeast Asian Nations (ASEAN) Region.

According to the 2022 report of the World Travel & Tourism Council (WTTC), Southeast Asia has been considered as a vibrant tourism hub with almost 140 million international tourist arrivals back in 2019. The industry has employed 41.8 million individuals or 13.2% of the region's total employment and contributed 11.7% to the ASEAN economy. Specifically, Cambodia, the Philippines, and Thailand were among the top countries in the region highly dependent on tourism in terms of total gross domestic product (GDP) contribution, constituting 25.8%, 22.5%, and 20.3% of their respective GDPs (WTTC, 2023). However, considering this major reliance of the region to the industry, the COVID-19 pandemic brought significant disruption and impacted different tourism sectors from the travel, tourism, to the hospitality sectors.

In the travel sector, per se, transportation and mobility has been considered as a major vector of pathogen distribution (Stavroulakis et al., 2021) which led for many economies within the region to temporarily cease global mobility through restrictions on the use of transportation, border closures, and shutdowns of major transportation systems from airlines, and trains, to public transportation. These lockdowns have resulted in a 49% decline in activity and a loss of close to $4.5 trillion (£3.7 trillion) compared with 2019 (Machiarelli, 2022) triggering a large decline in international tourist arrivals of 70% between January and October 2020 as compared to the same period in 2019 (UNWTO, 2020), and a huge decline in the travel industry profits of over $200 billion in Asia alone and a loss of $730 billion in export revenues globally (UNWTO, 2020). While there have been efforts to facilitate recovery by containing the virus and gradually easing restrictions, the persistent threat of higher global inflation has posed further challenges. This inflationary pressure has decreased household purchasing power, resulting in less pent-up savings. As incomes fail to keep pace with rising prices, there has been a noticeable stall in the travel demand (Machiarelli, 2022).

In the ASEAN region, the travel sector faces multiple challenges, as highlighted by Bowerman (2022) and Rodolfo (2022). Ongoing border closures in China, geopolitical tensions leading to higher jet fuel prices and airfares, European Air bottlenecks, concerns about the quality of airline services, and immigration issues are all contributing to disruptions in the industry. These factors collectively pose significant obstacles for the travel sector in the region, requiring careful attention and proactive measures to ensure its resilience and recovery.

The paralyzed travel sector has had far-reaching consequences, including stagnation in many tourisms business and decline in tourist receipts (Chen et al., 2021), decline in consumer demand, low cash reserves, shutdowns in the hospitality centers (Florida, 2020), and a general massive decline in tourist visits, resulting in severe impacts on the tourism and hospitality sectors due to their interconnected nature. In the tourism sector, economies, livelihood, public services, and opportunities have been drastically impacted by the shift in tourist arrivals and operational challenges brought by the lockdowns, travel bans, and imposed physical restrictions (Karabulut et al., 2020). Export revenues from international tourism have plunged 63% in 2020 and 61% in 2021 amounting to a combined loss of US$ 2.1 trillion in these 2 years (UNWTO, n.d.a). The huge decline has put millions of jobs and livelihoods at risk, where small businesses, women, youth, and the informal economy are at the biggest risk. While no nation was unaffected, destinations and economies that have been most reliant on tourism for job and economic growth have been hit the hardest including Small Island Developing States (SIDS), Least Developed Economies (LDCs), and African economies. The further decline of the sector has resulted in larger consequences including budget cuts on biodiversity conservation, a rise in poaching, looting, and in consumption of bushmeat due to decreased enforcement of tourism rules, pressure on heritage conservation, and postponement of many intangible cultural heritage practices such as traditional festivals and gatherings and closure of tangent markets (UNWTO, n.d.b).

In ASEAN, tourist numbers dramatically decreased by trifold compared to the prepandemic numbers (Alisya et al., 2021) as most economies pursued draconian measures to prevent the spread of the virus bearing a huge impact both socially and economically. While some economies have successfully kept their numbers down like in Thailand and Vietnam, other economies such as Indonesia, Malaysia, and the Philippines have constantly struggled to keep their cases in check and contain the virus leading to the prolonged spur of the virus (Vichit-Vadakan, 2021) which kept various tourist destinations temporarily halted for a longer period.

Although the pandemic has spurred new developments and innovations in the travel industry, such as the surge in teleconferencing and communication technologies, artificial intelligence, digital border management such as e-visas and e-gates, the concept of travel bubbles; revenge travel, a term used to describe a surge or boom in tourism following the COVID-19 pandemic (Liu & Wang, 2023); and the gradual recovery of destinations through domestic tourism and niche-based tourism, these developments cannot fully offset the significant losses incurred by the industry due to the decline in international tourism. While they have provided some relief and opportunities for recovery, the industry continues to grapple with the far-reaching impact of the reduction in international travelers, affecting not only the sector itself.

The closure of numerous destinations and the decline in tourist arrivals have devastated the hospitality sector, particularly due to its heightened vulnerability to health-related risks. This resulted in a sharp decline in hotel occupancy rates, reaching almost 50% in 2020. This led to the closure of hotel operations or their conversion into quarantine facilities, as observed in various economies, including the Philippines. Other establishments, such as restaurants and theme parks, also faced similar challenges. To cope with the situation, cost reduction measures were implemented, jobs were terminated, and loan repayment deferments were arranged to compensate employee salaries, leading to dismay in hotel stock performances and skill misalignment, reminiscent of the experience during the severe acute respiratory syndrome outbreak in 2003 (Kaushal & Srivastava, 2021). The hotel industry in Asia alone endured a staggering 57% drop in hotel occupancy rates, resulting in 63.4 million job losses and a revenue loss of US$1.041 million due to massive booking cancellations (Agustina & Yosintha, 2020). These statistics underscore the severe hardships faced by the hospitality sector in the wake of the pandemic.

As such, the COVID-19 pandemic shed light to numerous vulnerabilities within the tourism industry framework, notably its pronounced reliance on a destination-centric approach, where tourists' sole focus is to rely on their destination rather than the way to reach them and the other factors affecting their tourist experience. Although the destination-centric focus of the tourism industry provides tourists with immersive and authentic travel experiences and contributes to the economic growth of destinations, it also exposes certain drawbacks and gravely affects destinations in times of crisis like the COVID-19 pandemic. One prominent concern is the strain it places on local infrastructure development by overwhelming existing facilities such as roads and transportation options, water, and other natural and manmade resources due to sudden influx of tourists. This leads to worsened congestion, environmental degradation, and stretching of the current capacity of infrastructures beyond its limits.

Additionally, this approach can commodify intangible tourism assets by changing and treating culture and other assets as a touristic attraction or an economical commodity in general (Greenwood, 1978), which potentially undermines their cultural value. Focusing on the destination aspect (tourism) alone leads to insufficient efforts in enhancing other umbrella sectors, such as travel and hospitality, resulting in an uneven growth within the broader tourism ecosystem, which affects the overall sustainability of the industry.

Given this context, a paradigm shift becomes imperative in achieving a more sustainable new normal tourism industry. The post-COVID-19 era demands a comprehensive reevaluation of the TVC, necessitating a transformative approach to align the sector with forthcoming health and environmental challenges (Machiarelli, 2022). Although this has been considerably gaining attention in the academia, the focus of most literature has predominantly centered on developing resilient supply chains (Chowdhury et al., 2021; Dubey et al., 2021), stakeholder participation and governance (Roxas et al., 2020) and assessing of the current industry scenario (Brouder et al., 2021; Lew et al., 2020). Some literature also made speculations and recommendations for the rebuilding of industry based on present trends and mindset (Gössling et al., 2020; Gowreesunkar et al., 2023).

Exploring the intersection of these areas holds a huge potential to establish a more inclusive and integrated value chain. This entails establishing a harmonious and cooperative linear chain framework enhancing the end-to-end experiences for the new-normal tourists. However, to achieve such, it is necessary first to distinguish the travel, tourism, and hospitality sectors from one another as they are often get lumped together due to their interconnected and complementary elements. Avoiding interchange or lumping of their definitions in academic discourse is essential for a nuanced understanding of the post-COVID-19 tourism framework and guiding policy actions within these sectors.

Guided by this objective, the chapter builds on developing a novel paradigm for the TVC. The chapter commences through a deconstruction of the travel, tourism, and hospitality sectors – scrutinizing their definitions and functions while addressing overlaps. The chapter also revisits the TVC concept using Partale's model (2020), mapping its functional levels, and proposing enhancements for a sustainable tourism model in the better normal.

To address the concerns surrounding the present TVC, the chapter proposes strategies for establishing the new normal framework. This framework integrates discussed models, emphasizing sustainability while aligning with the current tourism. The chapter caps off by presenting some case studies from ASEAN destinations illustrating practical applications of the framework, enhancing the new normal tourist experience standards while prioritizing sustainability and inclusivity; and resilience against future crises like pandemics.

Deconstructing – Travel, Tourism, and Hospitality

> 'When I use a word,' Humpty Dumpty said in a rather scornful tone, 'it means just what I choose it to mean – neither more nor less.' – Lewis Carroll, *Through the Looking Glass and What Alice Found There* (p. 124).

This approach to the question of definition has been widely referenced in discourses relative to approaches to defining tourism. A few academics subject themselves to this Humpty Dumpty approach in defining this term.[1] While language is considered subjective and fluid, Heeley (1980) stresses the importance of having a "terminological exactitude" (p. 11) in defining this term. However, with tourism being a complex industry, deconstructing the various sectors, such as travel, tourism, and hospitality, can potentially expose overlapping boundaries

[1]Humpty Dumpty approach is a semantic approach where words are assigned with meanings based on the speaker's intention or personal interpretation rather than the established definition or shared convention – posing a problem in communication since this led to ambiguity and misunderstanding. It originated from Lewis Carroll's "Through the Looking-Glass"; however, its origin in the academic discussions cannot be traced and presumably coined by academics themselves.

and intricacies. This lack of clarity can lead to diverse interpretations and debates on what constitutes tourism and its various components.

Clawson and Knetsch (2013), Mitchell (1984), Baggio (2008), and Jakulin (2018) highlighted the complexity of the tourism industry and its socioeconomic system's deep interconnectedness, while Fennell (2015) discusses the intimate intertwinement of life's economic, sociocultural and environmental fabric. Disparities and ambiguities in economic and technical definitions further complicate delineating boundaries among the three sectors: travel, tourism, and hospitality. Hence, there is a need to disentangle these nuances and establish distinct demarcations. Recent efforts have aimed to simplify and define these terms more broadly.

According to Medlik (2012) from the Dictionary of Travel, Tourism, and Hospitality, travel refers to the most common activity for people, involving a journey from one place to another using varied mode of transport undertaken for leisure and any purpose by a range of individuals.

Tourism, however, is a more complex term. Scholars like Mathieson and Wall (1982), Fennell (2015), and Leiper (1979) have attempted to define it. Leiper (1979) views tourism as an open system consisting of five elements interacting with the larger environment. This definition appears to be like Mathieson and Wall (1982), which integrated three elements in defining tourism: (1) a dynamic element, which involves *traveling* to a specific destination, *staying*; and (3) *economic, social, and physical consequences of the above*. Fennell (2015) includes market, travel, destination, and marketing as part of the tourism system, although this can blur the line between travel and tourism in the absence of purpose of travel.

Consequently, the United Nations World Tourism Organization (UNWTO) defines tourism as "the movement of people to economies or places outside their usual environment for personal or business/professional purposes." For travel to be considered tourism, it must extend beyond 24 hours but not exceed 1 year in duration. Medlik (2012) simplifies this as "travel for pleasure with an overnight stay," (p. vii) highlighting that while all tourism involves travel, not all travel is tourism.

On the other hand, hospitality can be defined as the practice of being hospitable, which involves the reception and entertainment of guests with goodwill Wharton (2006). The Dictionary of Travel, Tourism, and Hospitality defines hospitality as providing accommodation, food, and drink for people away from home in exchange for a reward. Hospitality has been found crucial in the success of service institutions and tourism providers, distinguishing them and enhancing guest satisfaction (Walters et al., 2021).

By acknowledging the interconnection and overlap among travel, tourism, and hospitality sectors and adopting simplified definitions, their distinct roles in the global value chain are clarified.

All things considered, applying these concepts and ideas in assessing the effects of the COVID-19 pandemic in the industry plays a crucial part in the process of shifting paradigms. Reevaluating the existing value chain in these sectors is an imperative to disentangle potential confusions and guide appropriate actions needed. Historically, especially pandemic and prepandemic policies have focused predominantly on destination-centric actions, neglecting the equally vital aspects of travel and hospitality leading to inconsistencies in the overall tourist experience.

In the post-COVID-19 era, resilience in tourism is a paramount. Rebuilding the trust and confidence of travelers entails making travel safer and seamless, necessitating a transformation of the tourism framework, which involves devolution of different tourism functions and the value chain itself. The need to revisit the TVC is a necessity – rather than a choice.

Shifting Paradigms: Revisiting the TVC

Considering the number of players jointly meeting the tourism demand, tourism has been considered as a value chain. This concept, akin to value chains in the business sector, originated from Porter (1985), integrating competitive advantage into strategic decision making. Partale (2020) elaborated that value chains "encompass the value creation process of products or services, and on the other, a management approach that helps to achieve strategic objectives systematically." In tourism, value chains encompass interrelated activities within *a destination* that contribute to the entire tourist experience (Morales-Zamorano et al., 2020). However, applying the value chain concept can be complex due to the involvement of various stakeholders from different industries (travel, logistics, transportation, attractions, accommodations) and of different sizes (small to medium enterprises) (Goni & Yustika, 2019), as well as its susceptibility to economic, social, political, environmental, and technological uncertainties (Madera et al., 2017), social media, changing tourism demand, innovations in marketing, communications, and skill set due to its service-oriented nature.

To address these uncertainties, collaboration between the private sector, the public sector (government), and civil society is pivotal in the chain. This collaboration facilitates the access of resources, adds value through the various stakeholder processes, and delivers the end product (service) to consumers (tourists) integrating travel, tourism, and hospitality sectors. This systematic approach aids in identifying opportunities, enhancing stakeholder value, and striving for overall sustainability.

Various models and literature have visualized the linear nature of the value chain, depicting phases such as preconsumption (information search), consumption (during the trip), and postconsumption (recollection) experiences, among others (Akhoondnejad, 2015; Yilmaz & Bititci, 2006). These models highlight the importance of tourist satisfaction in establishing loyalty, competitive advantage, higher revisit rate, and revenue to the destination.

Despite the variations in the models, these literature all share a common depiction of the tourist journey from the origin to destination. However, within this straightforward framework lies a series of value-adding activities intricately connected through complex interactions among the actors at different levels, forming the *broader value creation system*. This not only maps the entire journey of the tourist, but also highlights interdependence and collaboration among stakeholders, elucidating the distinction between travel, tourism, and hospitality.

Furthermore, in addition to the segments of the linear value chain, the tourism value creation system also has functional levels (see Table 1.1) with distinct actors and functions necessary for achieving the objectives of the TVC (Partale, 2020).

The TVC encompasses a series of transactions spanning from a tourist's pre-trip experience, a trip experience, and the post-trip experience (Rahmiati et al., 2019); and understanding and redefining the chain requires evaluating different functional levels to align with current needs and demand of the industry.

The global TVC faced significant disruptions during the pandemic due to travel restrictions and safety interventions leading to the collapse in demand and business closures. Although macro-level components remained relatively stable, shifts occurred at the meso level impacting marketing, product development, and strategic planning brought by changes in consumer behavior patterns and the industry dynamics. Meanwhile, at the Micro level, functions and services under the travel, tourism, and hospitality sectors experienced shutdowns and closures, underscoring the interdependence between functional levels and sectors that exist even from prepandemic, leading to a domino and cascading effect across various tiers and domains during periods of disturbances like the pandemic.

Evidently, ASEAN, due to its proximity and economic links to China, where the initial outbreak of the novel coronavirus was declared a global health emergency, has experienced disruption of supply chains, decrease in travel, trade, and investment flows, due to implementation of different containment measures such as lockdowns and community quarantines. Local economies struggled further as restrictions on large-scale events, transportation, accommodation services, and leisure activities were imposed due to stringent policies and a lack of demand. Despite the different scale and scope of lockdown measures implemented across ASEAN economies, it slowed down and impaired economic activities, including the tourism sector, which is considered a significant source of income for these economies.

The combination of the economic turmoil, brought by sharp decline in oil prices, severe downward pressures on asset prices in both the equity and fixed income markets, decline in exports, especially in Vietnam and Thailand, alongside uncertain healthcare conditions due to the virus surge and new COVID-19 variants, exacerbated the tourism industry's sudden decline leading to business shutdowns, temporary closures of destinations, job loss and other socioeconomic downturns with more medium-term impacts on poverty and welfare particularly affecting vulnerable sectors such as women, youth, and low-income groups, mainly comprising the micro level (Organisation for Economic Co-operation and Development [OECD], n.d.).

At the meso and macro levels, the region has implemented policy responses to revive and support the tourism sector's recovery and resilience building, including digitalization, implementing travel bubbles, domestic and niche tourism promotions, integration of artificial intelligence, and newer approaches on marketing, communication, product development, and destination management.

The pandemic's impact on the global TVC, evident in industry income decline, job loss, and business shutdowns, necessitates a redefinition of the consumer journey in the postpandemic era. This involves restructuring the TVC model addressing health and environmental challenges while navigating the current

Table 1.1. The Functional Levels of the TVC.

Functional Level	Role in Tourism Value Creation	What is Included
Macro level	This comprises the fundamental elements of the offering of a destination and functions that are **prerequisites for tourism development in the region**. They indirectly **affect the value added to the region;** however, they must be put in place to enable the different business models to be developed in the region.	(1) **Natural and sociocultural resources of a region:** Landscape, climate, vegetation, architectural heritage, local customs, and much more besides (2) **Political and legal framework** (3) Tourism infrastructures (4) Executing agencies
Meso level	This level includes the overarching support functions critical for tourism enterprises and steering of tourism activities to achieve the objectives of the destination. Likely, they **indirectly affect the value added** but are deemed necessary in allowing business models to be developed successfully and achieving overarching objectives such as increasing quality of life and reducing environmental burdens.	(1) Strategic planning (2) Stakeholder management (3) Provision of offerings (4) Professionalization and quality management (5) Product development (6) Communications and marketing
Micro level	This level consists of all functions that create the direct value added through tourism. Companies and different actors are found at this level and directly affect **the tourism value added. The more items and services sold here, the more money stays in the region, hence more value added**. However, its success relies on how the **macro and meso levels are managed**.	This level includes all the functions and service providers in the industry, including: (1) Travel agencies and insurance (2) Transportation functions (all modes of travel) (3) Shuttle services (4) Accommodation (5) Subsistence and relevant experiences

(Continued)

Table 1.1. *(Continued)*

Functional Level	Role in Tourism Value Creation	What is Included
	This level directly shows the distinction among travel, tourism, and hospitality sectors through the actors present here. The functions and services at this level directly adhere to customer demands along their journey from travel planning and booking to getting there, getting around, and the other activities they can be involved in the region.	(6) Shopping facilities and options (7) Leisure activities and infrastructures (swimming pools, nature monuments, events, courses, trade fairs)

Source: Author's elaboration on Partale (2020); table constructed by the author.

economic landscape. This approach aims to strengthen each sector while fostering cooperative competition (coopetition), resilience, and sustainability across the travel, tourism, and hospitality sectors and the entire chain.

While empowering each sector to recuperate from their loss in the past years and build resilience for the future is important, integrating the value chain concept enormously aids in understanding the multitiered and complex value creation system, the different actors and concerned entities of the industry as well as pinpointing the possible entry points and areas of ineffectiveness where interventions can be made (Partale, 2020).

Based on the proposed model of Biancone et al. (2020) adapted from Porter's Value Chain Theory (1985) factors such as tourist attractiveness, receptive factors, new business models, digital infrastructure improvement, and culture are considered important for tourism management during the pandemic and future regeneration efforts.

However, while various frameworks have tackled the TVC, only a few addressed the need for a sustainable and integrated framework to adapt to the changing postpandemic landscape. This presents a promising yet complex area for research and destination planning and management, particularly within the context of the evolving value chain and tourism literature (Hjalager et al., 2016).

The New Normal TVC and the Future Ahead of It

Designing a comprehensive and integrated value chain for the new normal requires ardent work as it concerns different factors and dimensions (economic, environmental, and sociocultural) that cannot be approached with a one-size-fits-all plan, especially given the varying impact of the pandemic on different destinations. However, the author tried to develop standard recommendations that can be adapted and tailored to suit various destinations as summarized in Fig. 1.1.

(1) Strengthening participation across functional levels by integrating travel, tourism, and hospitality in one value chain and shifting from a destination-centric to a community-centered framework

Apart from the drastic effect of the pandemic on the travel, tourism, and hospitality sectors, the passage highlights the fragmented nature of the travel, tourism, and hospitality sectors which results to ineffective policy implementations, hiccups on end-to-end travel experience, and conflicting functional level participation. Bridging these gaps necessitates acknowledgment and the shifting of the current destination-centric paradigm to a community-centered point of view.

A community-centered approach maximizes benefits to the community, visitors and travelers, and culture while minimizing negative environmental and sociocultural impacts grounded on needs of the community, the tourists, and the stakeholders surrounding it.

Shifting paradigms demand enormous resources and restructuring, but enabling community and stakeholder participation from the three sectors and

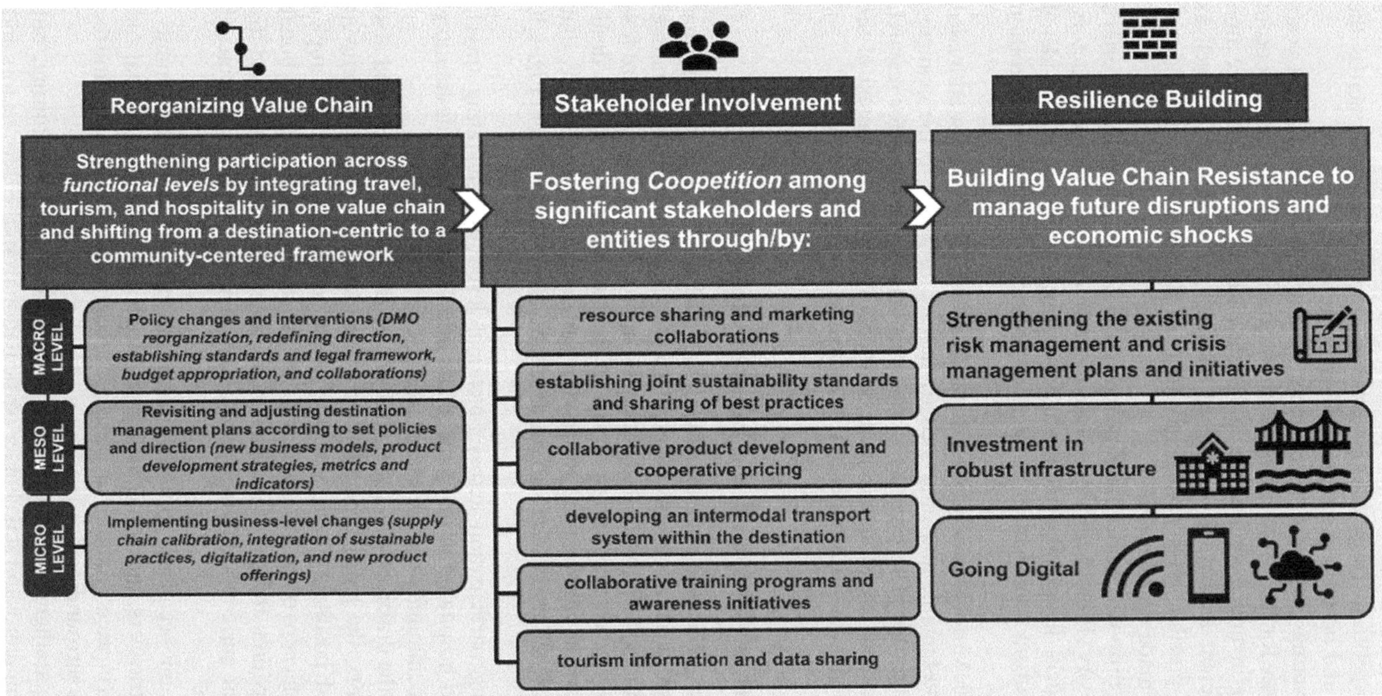

Fig. 1.1. Author Recommendations in Establishing the New Normal TVC. *Source:* Author's illustration.

even the local community can facilitate the transition. Opening communication and collaboration, through focal group discussions and consultations fosters trust and social cohesion. Assessing the entire value chain holistically reveals possible loopholes and flaws across exacerbated by the pandemic, such as the accessibility issues, quality and safety of mobility options, the vulnerability of MSMEs and other tourism sectors to shocks and disruptions which jeopardizes tourist journey.

Establishing tourism leadership inclusive of sectoral stakeholders down to the residents is crucial in addressing these concerns. Their involvement in the decision-making processes and designing strategies ensures more informed and context-specific action steps, and balanced policies ensuring sociocultural preservation, resistance, and flexibility to shocks and local economic benefit while advancing sustainability.

Strengthening the participation across functional levels (macro, meso, and micro) is essential in implementing concrete solutions. Each level bears responsibility for restructuring the current framework and transitioning the entire value chain toward a more comprehensive, holistic, and interconnected chain. Table 1.2 describes sample-specific interventions that can be done at each functional level:

Table 1.2. Sample Fundamental Policy Changes and Interventions That Functional Levels Can Adapt in Restructuring the Industry.

Functional Level	Intervention
Macro	**Policy changes and interventions:** (1) Restructure Destination Management Organizations (DMOs) and tourism leadership bodies to include sectoral representatives from the travel, tourism, and hospitality sectors, alongside community stakeholders. (2) Redefine tourism direction to align with the new paradigm and the integration of the three sectors. (3) Integrate sustainable tourism standards and certifications such as GSTC, alignment with UN SDGs, and incentivize responsible tourism practices. (4) Strengthen legal framework to penalize environmentally destructive practices and prioritize cultural preservation toward tourism development. (5) Appropriate budget for: • People-centric tourism infrastructures; • Upskilling and capacity-building initiatives; • Public health, insurance, risk mitigation measures;

(Continued)

Table 1.2. *(Continued)*

Functional Level	Intervention
	• Industry pandemic recovery efforts; • Research and promotion of new and lesser explored destinations; • Environmental protection and management initiatives • Cultural preservation programs; and • Digitalization (ICT infrastructures). (6) Financial support and access to capital for MSMEs. (7) Facilitate more efficient public–private partnerships among community enterprises and potential investors. (8) Forge collaborations with nongovernmental organizations (NGOs) and development agencies.
Meso	**Revisiting and adjusting destination management plans according to set policies and direction, which includes re/developing the following areas:** • Integrating community engagement (esp. MSMEs) and sustainability in new business models. • Marketing and promotional strategies, especially communication of the direction to concerned stakeholders. • Stakeholder management mechanisms. • Crisis and emergency management strategies. • Metrics on monitoring and evaluating the social, economic, and environmental impacts of initiatives and development activities. • Innovative tourism practices and product development strategies. • Supply chain and procurement management. • Tourism standards and professional certifications. • More effective data collection, benchmarking, and feedback mechanisms for community and visitors to provide input, share their experiences in the destination, and suggest improvements.
Micro	**Implement business-level changes**, which include: • Sourcing services and goods from local and sustainable providers and fostering collaboration with them under fair terms and conditions. • Integrating sustainable and green practices in hospitality (e.g., wastewater recycling, energy conservation), MSMEs, and transport providers (e.g., turning to greener fuels, creating bicycle rentals, or investing in electric shuttles).

Table 1.2. *(Continued)*

Functional Level	Intervention
	• Investing in low-impact and low-footprint tourism offerings. • Incorporating sustainability as part of employee capacity building. • Promoting responsible travel behaviors throughout the tourist journey. • Investing in digitalization to enhance tourist experience by integrating contactless services, automated visitor identification, interconnected booking systems, and utilizing a unified digital platform to assist tourists from booking transportation and accommodation to crafting their travel itineraries. • Integrating health and safety measures into business practices and strengthening the imposition of public health standards. • Creating curated packages for different tourist segments and rising markets such as off-peak travelers, workations, wellness tourism, and adventure tourism.

Source: Author's suggestion and elaboration on model of Partale (2020).

In general, each functional level plays a crucial role in ensuring a smooth and seamless transition to the new normal.

At the macro level, policy interventions remain crucial in setting the direction for the value chain, creating an enabling environment for the paradigm shift While the macro level's impact on direct value addition may be indirect, its role in shaping the overall direction is indispensable.

Meanwhile, the meso level reinforces and shapes the direction and policies set focusing on the essential functions such as upscaling the human capital, communicating the vision s to the entire community, situating stakeholders in the strategic plan and offering resources to translate plans to actionable initiatives and mitigating adverse effects.

The micro level, on the other hand, translates the direction to experience where the value added to the community, particularly on each of the sectors linked under the value chain, heavily depends.

The interventions shown illustrate how the proposed integrated value chain connects the functional levels and the different sectors of the industry. By linking the three sectors and empowering functional levels to collaborate, the value chain becomes more sustainable and contributes to poverty reduction while preserving environmental and sociocultural dimensions which are essential foundations in pursuing sustainable tourism in the new normal.

(2) Fostering coopetition among significant stakeholders and entities
Once the destination has established a clear direction toward sustainability and an enabling environment conducive to the shift, enabling coopetition across all stakeholders becomes crucial. Coopetition encourages collaboration among various stakeholders to achieve mutual benefits while allowing healthier competition in the market, advancing sustainable practices with long-lasting impacts. Dagnino (2009, p. 29) further defined coopetition as "within inter-firm interdependence, both processes of value creation and value sharing take place, giving rise to a partially convergent interest (and goal) structure, where both competitive and cooperative issues are simultaneously present and strictly interconnected." When managed properly, this leads to creation of sustainable advantages among involved sectors (Chim-Miki et al., 2023a; Della Corte & Aria, 2016), and social value generation, apart from balancing the negative effect of competition in tourism (Chim-Miki et al., 2023b).

In the context of sustainable tourism, coopetition can be present in several ways:

- *Resource sharing and marketing collaborations:* Key tourism players share and maintain resources like tourism infrastructure, transportation options, renewable energy sources, and waste management technology; promoting these practices through joint marketing efforts, which leads to reduced costs and footprint and improved efficiency.
- *Joint sustainability standards and best practices:* Hospitality sectors can craft localized sustainability framework applicable to each entity within the destination and share best practices in waste reduction, energy efficiency, water conservation, and other related dimensions aligned with sustainability goals.
- *Collaborative product development and cooperative pricing:* Tourism enterprises can collaborate in creating new tourism products, such as tour packages that connect tour operators and agencies to hotels and other local businesses, which can also expand their market. Discounts and pricing strategies can also be applied to tourists availing of these offerings.
- *Intermodal transport system within the destination:* The transport sector can establish an integrated and sustainable transport network that taps public transport networks, walking, and cycling which reduces individual car usage and ensuring seamless mobility within destination through quality tourism transport infrastructure.
- *Collaborative training and awareness initiatives:* Enterprises can cosponsor capacity-building initiatives for businesses and communities within the region to standardize tourism practices and upskill sectors on the needs of the new normal tourism, including digitization within the area. They can also organize sustainability-focused initiatives such as conferences, symposiums, or responsibility statements to bring together

different stakeholders, including visitors, and align them with the direction set at the macro level.

- *Tourism information and data sharing:* Competing businesses and entities such as the government and tourism organizations can share data on tourism impacts such as visitor numbers, resource consumption, visitor evaluation, and feedback to collectively address sustainability challenges and forecast potential travel trends, enhancing resource management and proactive initiatives.

These are just some initiatives that tap the potential of coopetition to advance sustainable tourism while considering the paradigm shift. The concept of linking the three sectors of travel, tourism, and hospitality embodies coopetition as shown in Fig. 1.2.

Coopetition comes with challenges, such as conflict of interest, information sharing concerns, reconciling cooperation and competition, and the risk of exploitation through power dependence and opportunism. However, by establishing clear rules, transparent communication, dispute resolution mechanisms, and defined boundaries, coopetition can succeed, reap benefits for stakeholders, and contribute to empowering destinations to become more sustainable in the long run.

(3) Building value chain resistance to manage future disruptions and economic shocks

COVID-19 is just one example of the crises and shocks the world could encounter. The ASEAN region has always been prone to climate change-related disasters (extreme weather conditions such as typhoons, excessive flooding, extreme heat, and drought), economic turmoil, and other

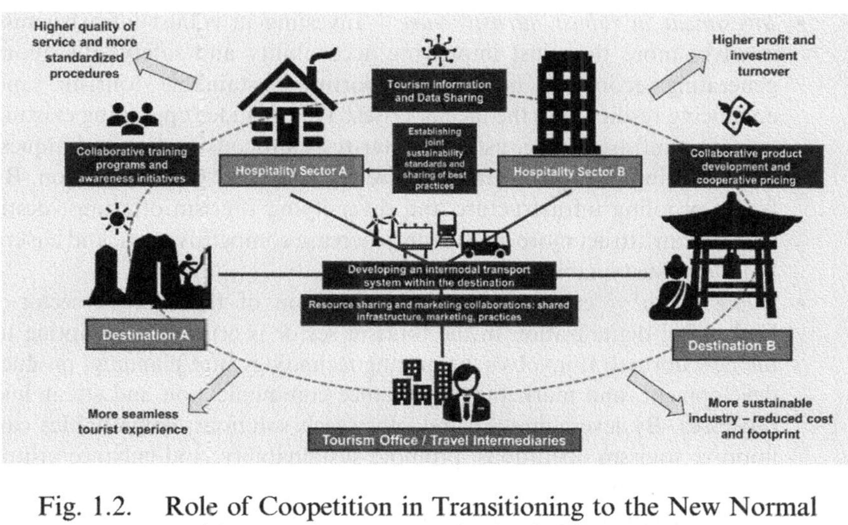

Fig. 1.2. Role of Coopetition in Transitioning to the New Normal and its Benefits. *Source:* Author's illustration.

human-caused crises. The pandemic was instead a reminder of how vulnerable the industry is to these disruptions, emphasizing the importance of building long-term resilience and effective crisis management in the shift toward a community-centric tourism framework and adapting to the new normal.

Implementing the recommended actions outlined in the previous sections will significantly contribute to instituting resilience and crisis preparedness efforts across the three sectors and recovery from the impacts of the pandemic and other shocks. These actions can be further reinforced by the following recommendations, which highlight the importance of building proactiveness across the destination and its stakeholders:

- *Strengthening the existing risk management and crisis management plans and initiatives* – Every destination has its own risk and crisis management plans. However, most existing frameworks must be updated and have minimal provisions for crises like a pandemic.
  By creating provisions such as:
  - improved crisis communication and response and evacuation procedures;
  - conducting readiness and tourism impact assessments;
  - bolstering emergency financing for tourism stakeholders, especially for MSMEs, through an allocated resilience fund;
  - comprehensive monitoring system and reporting; and
  - continuous collaboration integration of different stakeholders from the planning to the implementation stage,
    we can address the gaps in the existing crisis management and tailor the response to the future and specific needs of the community in the face of these crises.
- *Investment in robust infrastructure* – Investing in robust infrastructure involves more than just improving accessibility and safety. It's about generating economic benefits, supporting sustainable tourism, and enhancing resilience in the face of crises. This includes upgrading existing transport infrastructure, using disaster-resistant construction techniques, and investing in digital infrastructure for smoother communication. By future-proofing infrastructure and diversifying tourism offerings, destinations can attract more investment, increase competitiveness, and ensure long-term sustainability.
- *Going digital* – emphasis on the digitization of the tourism sector – Embracing digitalization in the tourism sector is crucial for adapting to the new normal. It involves integrating technology into planning, product development, and marketing to enhance communication and streamline processes. By leveraging AI and other tech solutions, communities can improve tourism operations, promote sustainability, and enhance visitor experiences. This shift toward digitalization not only improves efficiency but also contributes to long-term environmental conservation and resilience in the tourism industry.

These strategies can be integrated with the action steps mentioned above to guarantee the promotion and sustenance of resilience as the industry transitions into the new framework and adapts to the new normal. The following section discusses how these strategies are already evident in the recovery efforts of select destinations within the ASEAN region.

Case Studies in ASEAN: Successful Examples From Thailand, Malaysia, and Indonesia

Surprisingly, the key components of the proposed framework, shifting toward a community-centric and sector-integrated approach, have been actively implemented in select destinations within the ASEAN Region, showcasing notable success stories in Thailand, Malaysia, and Indonesia. These strategies, highlighted by Channel NewsAsia (CNA) Insider (2022), emphasize sustainable tourism principles and have proven effective in adapting to the new normal and sustaining the industry amid challenging circumstances.

- *Thailand: Koh Tao Island – a case of community building*

Dubbed the "Turtle Island," Koh Tao Island is a picturesque island in the heart of the Gulf of Thailand. It is known for its perfect white sand beach, sideline activities such as rock climbing, and diverse marine biodiversity with species including whale sharks and rays, turtles, and tropical coral reefs (Reeves, n.d.), making it a popular diving spot, hosting over 600,000 annual visitors. As annual arrivals increased, its palm-fringed beaches were built with villas, hotels, and several establishments, which drove economic progress on the island, especially in the local community. However, while development was underway, over-tourism on the island extensively damaged coral reefs and spurred environmental degradation, affecting the community and their natural balance (CNA Insider, 2022). Although this issue of over tourism in Koh Tao Island is a similar case to that of Maya Bay in Krabi Province, Thailand (deterioration of coral reefs, overcrowding which led to disastrous degradation of coastal resources) (Koh & Fakfare, 2020; Pathak, 2024), which led to its temporary shutdown, the fate of Koh Tao Island diverges as the pandemic also took a huge blow in the island.

The pandemic came, and while the temporary shutdown gave the island a breather, this has created an economic shock as it halted operations and income flow to its households. However, despite lost income, the local government and conservation activists encouraged the enterprises and community to unite and work toward achieving sustainability on the island. At the same time, the industry is still shut down. This includes balancing economic leverage and community sustainability to ensure consistent and regular tourist arrivals rather than short-term peaks.

To achieve this goal, the community chipped in to conserve and preserve the nature of Koh Tao by pioneering initiatives such as regrowing the corals supplemented by the implementation of visual coral courses to help people identify dead or young corals. Enterprises like community resorts have also committed to recycling and conservation efforts to alleviate the concern of the decreasing space to hold rubbish and charging environmental fees of 20 Baht or $0.50 for every visitor, expected to yield $80,000 to $300,000 (UNDP Biodiversity Finance Initiative, 2022).

Acknowledging the economic impact of the temporary shutdown on several vulnerable groups like fishermen and service providers, crowdfunding campaigns were launched to deliver basic income to these groups currently working under cash-for-work programs like debris cleaning on the island. Cash-for-work programs are considered as appropriate crisis and safety net response (Andrews et al., 2012) which have been found beneficial not only as an economic aid but a way to benefit social environment in regions affected by the pandemic and a way to encourage purchasing power and grow the existing markets of small medium enterprises (SMEs) (Fitriani, 2020).

In the long run, to ensure its sustainability, the island targeted diversifying its offerings, making it renowned for "all things natural," where the community will be essential in achieving this transition to a more sustainable Koh Tao island. This gesture has become a role model initiative for other islands in Thailand, emphasizing the critical role of the community in transitioning to a community-focused tourism framework, where there is a balance between economic growth and environmental sustainability and where all sectors – from the travel, tourism, and hospitality sectors work hand in hand in achieving this goal.

- *Malaysia: Willow Tree Campsite – cultural preservation is important too*

The new normal tourism saw a rise in visits in natural tourist destinations as these have been assumed to be safe and have the lowest risk of virus transmission (Ananda & Novianti, 2021) due to individuals' ability to maintain social distance and its outdoor setting (Craig & Karabas, 2021; Gössling et al., 2020). Among the popular outdoor tourism activities done across countries is camping which has been considered as a rising tourism segment with an over $150 billion profit (Rice et al., 2019).

In the ASEAN region, natural off-grid camping has always been a popular activity in Malaysia due to its environment suitable for this kind of activity, such as forests, hills, and mountain areas (CNA Insider, 2022). The pandemic has even spurred an increase in off-grid campers, encouraging greater reliance on outdoor activities and appreciation for the environment. One known campsite offering off-grid camping in Malaysia is the Willow Tree Campsites. Scattered across the country, the campsite is known to offer not just high-quality glamping and camping experiences but also sustainable practices.

Some practices include using nonfossil fuel in pumping water, using solar energy to light the camp, sourcing campfire wood from a licensed sustainable harvester, and teaching tourists about different recycling practices, such as food waste to compost. Sustainability is not only what the campsite does, but its guests are also practicing it (CNA InsiderFeb. 16).

While the pandemic helped many economies to discover new and exciting ecotourism activities in many destinations, it also drove these destinations to further appreciate the importance of cultural preservation as an integral mark of sustainability. Particularly, in Willow Tree, this has been the integral role of the indigenous community in the process – to help in the cultural preservation efforts. An authentic intangible cultural heritage (ICH) has been pertained as an important sustainable tourism resource, however, to achieve such local community and visitor awareness and institutional empowerment should be present and practiced in the destination (Kim et al., 2021).

Thus, in Tamwan community, campers are being exposed to community immersion activities apart from staying in bamboo-made chalets. They believe in the importance of cultural preservation as a parcel of sustainable tourism. This includes preserving heritage buildings, community practices, traditions, and their skills – and this has been the campsite's continuous commitment to its visitors. This emphasized how environmental preservation goes hand in hand with cultural preservation, as both of them play an essential role in committing to building a more sustainable tourism framework not only for the present visitors but, most importantly, for the future.

- *Indonesia: Batulayang Tourist Village – tourism numbers are not the end of all*

Tourism numbers are considered vital metrics with a direct impact on a destination's economic, social, and cultural aspects. These numbers are being used to assess industry performance, make informed decisions, and shape policies to support the industry's sustainable growth. Before the pandemic, this was the primary motivation of many economies and destinations, such as Indonesian tourist villages.

Tourist villages are considered as common sight in Indonesia. Established as a priority program by the Indonesian government to develop communities, tourist villages are considered as one of the growing models of sustainable tourism in the archipelago (Pickel-Chevalier et al., 2022). According to a content analysis study done by Sari and Lestari (2021) over 464 tourist reviews on TripAdvisor, it was revealed that tourist villages largely bring visitors to the archipelago due to its scenic landscape, traditional building and way of life, and overall village atmosphere. And among the approximately 1,200 tourist villages in Indonesia (Ministry of Tourism, n.d.), Batulayang Tourist Villages stands out as one of the most visited in the archipelago.

Located in West Java, Indonesia, the development of tourist villages such as the Batulayang Tourist Villages proved there is more to achieving the numbers. Made up of 22 neighborhood units, Batulayang offers a wide array of outbound activities for its visitors, including ATV, glamping, swimming activities, etc. (CNA Insider, 2022). But more than just these activities and features, the tourist village has been an exemplary example of how the community plays a vital role in preserving tourist destinations. Village residents (at least 75% of the 300 families) work shifts to ensure the visitors' well-being. Over the years, they made over $100000 worth of profit, which supported the lives of the community members.

However, when the pandemic struck in 2020, there was a significant drop in tourist visits to the village, which temporarily halted its operations and paused its income stream. Despite its promising business model, tourist villages in Indonesia are prone to difficulties due to "limited funds and absence of good tourism management" practices (Alawi et al., 2021, p. 90). These disruptions inhibit the development and sustainability of these tourist villages.

However, this loss in income and tourist visitors did not discourage the village from sustaining the industry, as this even pushed them to innovate and work beyond the numbers. Even though the homestays were suspended, the women upskilled themselves by learning crochet and cross stitch and selling them online through artificial plants, tea sets, moss connectors, and water bottle sleeves. They also learned and were taught upcycling techniques, which helped them turn coffee and detergent packaging into mats and accessories sold in the market (CNA Insider, 2022).

These activities may need to level the profit they got from their homestays. However, these helped them diversify their offerings and work on achieving a sustainable source of income for themselves. This initiative in Batulayang even became a model for the Indonesian government in building more tourism villages to improve the welfare of many of their communities, help them manage their resources, and be more economically independent and self-sufficient. Tourist villages have the potential to bring more entrepreneurial opportunities to the community while fostering environmental education and social disaster management (Wibowo et al., 2021). Through this, villages can solve the complex issue of unemployment within these villages while achieving sustainable tourism and sustainability in general – an achievement that no number can even define or measure.

These examples from Thailand, Malaysia, and Indonesia only show how shifting paradigms is not an impossible path toward sustainable tourism. The featured destinations have shown how the pandemic became instrumental in shifting priorities from merely earning tourism income and achieving tourist arrivals to involving the community in environmental and cultural preservation on the sidelines of innovation and development. These elements were found to be crucial in the suggested action plan above. They can be replicated by other destinations and regions within the region and across the globe.

Conclusion

The pandemic, over 3 years in, has exposed the critical disjoint nature of the global TVC. Through the years, due to its economic potential, the industry has reaped profit from excessive mass and destination-centric tourism at the expense of environmental degradation and community disempowerment.

While we do not undermine the efforts made to improve tourist destinations, the consequences of the persistent mass tourism spurred by the pandemic expose the need to pursue a more sustainable tourism industry by reconnecting the aspects of travel, tourism, and hospitality. Considering the breadth of this problem in the industry, in this chapter, the author attempted to "shift paradigms" by proposing a value chain framework which heeds to the needs of the new normal tourism. In formulating the foundation for the framework, the author deconstructed the travel, tourism, and hospitality sectors to identify overlapping functions and streamline approaches relevant to each sector. Drawing from Partale's model of TVC (2020), the author proposed concrete and comprehensive industry-based action steps geared toward reorganizing the value chain, enhancing stakeholder involvement, and resilience building.

Specifically, the recommended framework includes addressing potential gaps between the sectors, promoting active participation and cooperation (including coopetition) among stakeholders such as the community, private and public enterprises, and the different functional levels; and restructuring their functions adjacent to the innovations and adjustments heeding to the needs of the new normal transition. To demonstrate the efficacy of the framework, the author presented case studies from selected ASEAN destinations showcasing how these principles have been applied in recent years. This illustrated the potential of the framework to drive positive change and its adaptability across diverse tourism destinations within the region. Most importantly, this serves as a pivotal launchpad in steering the entire industry toward the direction of sustainability and inclusivity in the new normal.

Clearly, the pandemic has drastically changed the global tourism landscape. Beyond the challenges, it has exposed the need to restructure our perspectives toward the industry as a key to transitioning to the new normal. Changing paradigms aligned with the demands of the postpandemic era entails not only reconfiguring the past but also challenging archetypes, redefining the industry, and fostering collaboration. Embracing these principles and elements provides us with the compass to rebuild the tourism industry, aiming not just for safety, equity, and climate friendliness – but for a better industry for a better normal to come.

References

Agustina, I. T., & Yosintha, R. (2020). The impact of COVID-19 on hotel industry in Asian countries. *Jurnal Kepariwisataan Indonesia: Jurnal Penelitian Dan Pengembangan Kepariwisataan Indonesia, 14*(2), 159–167.

Akhoondnejad, A. (2015). Analyzing the pre-travel, on-travel, and post-travel behaviors of Iran's first-time visitors. *Journal of Travel & Tourism Marketing, 32*(8), 1023–1033.

Alawi, M. I., Habibi, M. M., Rochmadi, N. W., & Wiwoho, B. S. (2021). The problems of developing a tourist village. In *Empowering civil society in the industrial revolution 4.0* (pp. 90–94). Routledge.

Alisya, D., Gustiandza, M., & Rizky, K. (2021). *COVID-19 effects on international tourism in ASEAN economies*. ArcGIS StoryMaps. https://storymaps.arcgis.com/stories/d95fd479bc78438c9a63adcce65eacd5

Ananda, A., & Novianti, S. (2021). Tourists interest in camping as an alternative tourism activity during COVID-19 pandemic: The AIDA model application. *Journal of Tourism Sustainability*, *1*(2), 74–81.

Andrews, C., Ovadiya, M., Ribes Ros, C., & Wodon, Q. (2012). Cash for work in Sierra Leone: A case study on the design and implementation of a safety net in response to a crisis. https://documents1.worldbank.org/curated/en/321381468166178112/pdf/NonAsciiFileName0.pdf

Baggio, R. (2008). Symptoms of complexity in a tourism system. In *Tourism analysis* (1st ed., pp. 1–20). Cognizant Communication Corporation. https://doi.org/10.3727/108354208784548797

Biancone, P., Secinaro, S. F., Brescia, V., & Calandra, D. (2020). Employing value chain theory to address COVID-19 outbreak in tourism management: A resilience and stakeholder view. *International Journal of Business Research and Management*, *11*(3), 39–64.

Bowerman, G. (2022). *ASEAN Air Travel and Tourism POST-COVID 19* [Webinar]. FACTS Asia. https://www.factsasia.org/blog/webinar-asean-air-travel-and-tourism-post-covid-19

Brouder, P., Teoh, S., Salazar, N. B., Mostafanezhad, M., Pung, J. M., Lapointe, D., Higgins-Desbiolles, F., Haywood, M., Hall, C., & Clausen, H. B. (2021). Conclusions: Reflections and discussions: Tourism matters in the new normal post COVID-19. In *Global tourism and COVID-19* (pp. 281–292). Routledge.

Chakraborty, A., & Kar, A. K. (2021). How did COVID-19 impact working professionals – A typology of impacts focused on education sector. *The International Journal of Information and Learning Technology*, *38*(3), 273–282. https://doi.org/10.1108/IJILT-06-2020-0125

Chen, X., Duan, Y., Ali, L., Duan, Y., & Ryu, K. (2021). Understanding consumer travel behavior during Covid-19. *Sustainability*, *13*(23), 13330.

Chim-Miki, A. F., da Costa, R. A., & Oliveira-Ribeiro, R. (2023a). Dimensions of tourism coopetition for social value creation. *Anatolia*, 1–13.

Chim-Miki, A. F., da Costa, R. A., & Oliveira-Ribeiro, R. (2023b). Tourism coopetition for a better world: A cycle of creation, appropriation, and devolution of social value. *Current Issues in Tourism*, 1–15.

Chowdhury, P., Paul, S. K., Kaisar, S., & Moktadir, M. A. (2021). COVID-19 pandemic related supply chain studies: A systematic review. *Transportation Research Part E: Logistics and Transportation Review*, *148*, 102271. https://doi.org/10.1016/j.tre.2021.102271

Clawson, M., & Knetsch, J. L. (2013). *Economics of outdoor recreation*. Routledge.

CNA Insider. (2022, February 16). The push for sustainable tourism in Southeast Asia amid pandemic | CNA correspondent [Video]. YouTube. https://www.youtube.com/watch?v=xX5XcqQhlI8

Craig, C. A., & Karabas, I. (2021). Glamping after the coronavirus pandemic. *Tourism and Hospitality Research*, *21*(2), 251–256. https://doi.org/10.1177/1467358421993864

Dagnino, G. B. (2009). *Coopetition strategy: A new kind of interfirm dynamics for value creation. Coopetition strategy* (pp. 45–63). Routledge.

Della Corte, V., & Aria, M. (2016). Coopetition and sustainable competitive advantage. The case of tourist destinations. *Tourism Management*, *54*, 524–540. https://doi.org/10.1016/j.tourman.2015.12.009

Dubey, R., Bryde, D. J., Foropon, C., Tiwari, M., & Gunasekaran, A. (2021). How frugal innovation shape global sustainable supply chains during the pandemic crisis: Lessons from the COVID-19. *Supply Chain Management: An International Journal*, *27*(2), 295–311. https://doi.org/10.1108/scm-02-2021-0071

Fennell, D. (2015). Ethics in Tourism. In G. Moscardo & P. Benckendorff (Eds.), *Education for sustainability in tourism: A handbook of processes, resources, and strategies* (pp. 45–57). Springer. https://doi.org/10.1007/978-3-662-47470-9

Fitriani, A. (2020). The effectiveness of cash for work in handling the impact of COVID-19. *International Conference of Zakat* (pp. 127–138). https://doi.org/10.37706/iconz.2020.235

Florida, R. (2020, April 4). The geography of Coronavirus. *Bloomberg*. https://www.bloomberg.com/news/articles/2020-04-03/what-we-know-about-density-and-covid-19-s-spread

Goni, J. I., & Yustika, B. P. (2019). The presence of global value chain in coastal marine tourism. *Jurnal Riset Manajemen Dan Bisnis (JRMB) Fakultas Ekonomi UNIAT*, *4*(1), 137–152.

Gössling, S., Scott, D., & Hall, C. M. (2020). Pandemics, tourism, and global change: A rapid assessment of COVID-19. *Journal of Sustainable Tourism*, *29*(1), 1–20.

Gowreesunkar, V. G., Maingi, S. W., Roy, H., & Micera, R. (2023). Rebuilding tourism post pandemic–policy recommendations from global case studies. *Emerald Open Research*, *1*(13).

Greenwood, D. J. (1978). Culture by the pound: An anthropological perspective on tourism as cultural commoditization. In *Hosts and guests: The anthropology of tourism*. University of Pennsylvania Press.

Heeley, J. (1980). The definition of tourism in Great Britain: Does terminological confusion have to rule? *Tourist Review*, *35*(2), 11–14.

Hjalager, A. M., Tervo-Kankare, K., & Tuohino, A. (2016). TVCs revisited and applied to rural well-being tourism. *Tourism Planning & Development*, *13*(4), 379–395.

Jakulin, T. (2018). Systems approach to tourism: A methodology for defining complex tourism system. *Organizacija: Journal of Management, Informatics and Human Resources*, *50*(3).

Karabulut, G., Bilgin, M. H., Demir, E., & Doker, A. C. (2020). How pandemics affect tourism: International evidence. *Annals of Tourism Research*, *84*, 102991.

Kaushal, V., & Srivastava, S. (2021). Hospitality and tourism industry amid COVID-19 pandemic: Perspectives on challenges and learnings from India. *International Journal of Hospitality Management*, *92*, 102707.

Kim, S., Whitford, M., & Arcodia, C. (2021). Development of intangible cultural heritage as a sustainable tourism resource: The intangible cultural heritage practitioners' perspectives. In *Authenticity and authentication of heritage* (pp. 34–47). Routledge.

Koh, E., & Fakfare, P. (2020). Overcoming "over-tourism": The closure of Maya Bay. *International Journal of Tourism Cities*, *6*(2), 279–296. https://doi.org/10.1108/IJTC-02-2019-0023

Lew, A. A., Cheer, J. M., Haywood, M., Brouder, P., & Salazar, N. B. (2020). Visions of travel and tourism after the global COVID-19 transformation of 2020. *Tourism Geographies*, *22*(3), 455–466.

Leiper, N. (1979). The framework of tourism: Towards a definition of tourism, tourist, and the tourist industry. *Annals of Tourism Research*, *6*(4), 390−407. https://doi.org/10.1016/0160-7383(79)90003-3

Liu, X., & Wang, W. (2023). Revenge travel: Fact or myth? *Current Issues in Tourism*, 1–7.

Machiarelli, C. (2022, August 30). Update: How is Covid-19 affecting international travel and tourism? *Economics Observatory*. https://www.economicsobservatory.com/update-how-is-covid-19-affecting-international-travel-and-tourism

Madera, J. M., Dawson, M., Guchait, P., & Belarmino, A. M. (2017). Strategic human resources management research in hospitality and tourism: A review of current literature and suggestions for the future. *International Journal of Contemporary Hospitality Management*, *29*(1), 48–67.

Mathieson, A., & Wall, G. (1982). *Tourism: Economic, physical, and social impacts*. Prentice Hall.

Medlik, S. (2012). *Dictionary of travel, tourism, and hospitality*. Routledge.

Ministry of Tourism. (n.d.). *Tourist villages*. Desa Wisata. https://www.indonesia.travel/desawisata/en/home

Mitchell, L. S. (1984). Tourism research in the United States: A geographic perspective. *GeoJournal*, *9*(1), 5–15. https://doi.org/10.1007/bf00518313

Morales-Zamorano, L. A., Camacho-García, A. L., Bustamante-Valenzuela, A. C., Cuevas-Merecías, I., & Suarez-Hernández, Á. M. (2020). Value chain for agri-tourism products. *Open Agriculture*, *5*(1), 768–777. https://doi.org/10.1515/opag-2020-0069

OECD. (n.d.). *COVID-19 crisis response in ASEAN member states*. https://www.oecd.org/coronavirus/policy-responses/covid-19-crisis-response-in-asean-member-states-02f828a2/

Partale, K. (2020). *The TVC: Analysis and practical approaches for development cooperation projects*. Deutsche Gesellschaft für Internationale Zusammenarbeit (GIZ).

Pathak, A. K. (2024). Prospective tourism boom and the peril of overtourism: What lakshadweep can learn from Maya Bay. *Preprints*. https://doi.org/10.20944/preprints202401.1341.v2

Pickel-Chevalier, S., Bendesa, I. K. G., & Putra, I. N. D. (2022). The integrated touristic villages: An Indonesian model of sustainable tourism? In *Island tourism sustainability and resiliency* (pp. 262–286). Routledge.

Porter, M. E. (1985). *Competitive advantage: Creating and sustaining superior performance*. The Free Press.

Rahmiati, F., Ismail, Y., Amin, G., Simatupang, T., Larson, D., & Othman, N. A. (2019). TVC activities model for competitive advantage measurement. In *International conference on trade 2019 (ICOT 2019)* (pp. 43–47). Atlantis Press.

Reeves, B. (n.d.). 10 best things to do in koh tao: What is koh tao most famous for? Hotels.com. https://ph.hotels.com/go/thailand/best-koh-tao-things-to-do

Rice, W. L., Park, S. Y., Pan, B., & Newman, P. (2019). Forecasting campground demand in US national parks. *Annals of Tourism Research*, *75*, 424–438.

Rodolfo, C. (2022). *ASEAN Air Travel and Tourism POST-COVID 19* [Webinar]. FACTS Asia. https://www.factsasia.org/blog/webinar-asean-air-travel-and-tourism-post-covid-19

Roxas, F. M., Rivera, J. P., & Gutierrez, E. L. (2020). Mapping stakeholders' roles in governing sustainable tourism destinations. *Journal of Hospitality and Tourism Management*, *45*, 387–398. https://doi.org/10.1016/j.jhtm.2020.09.005

Sari, P. I., & Lestari, Y. D. (2021). Determinants of tourist satisfaction and dissatisfaction on tourism village. *Jurnal Pendidikan Ekonomi Dan Bisnis (JPEB)*, *9*(1), 9–24. https://doi.org/10.21009/009.1.2

Stavroulakis, P. J., Tzora, V. A., Riza, E., & Papadimitriou, S. (2021). Transportation, the pathogen vector to rule them all: Evidence from the recent coronavirus pandemic. *Journal of Transport & Health*, *22*. https://doi.org/10.1016/j.jth.2021.101087

Tasnim, Z., Shareef, M. A., Dwivedi, Y. K., Kumar, U., Kumar, V., Malik, F. T., & Raman, R. (2022). Tourism sustainability during COVID-19: Developing value chain resilience. *Operations Management Research*, *16*(1), 391–407. https://doi.org/10.1007/s12063-021-00250-8

UNDP Biodiversity Finance Initiative. (2022, December 8). *Better together in koh tao, Thailand.* https://www.biofin.org/news-and-media/better-together-koh-tao-thailand

UNWTO. (2020, October 27). *International tourism down 70% as travel restrictions impact all regions.* https://www.unwto.org/news/international-tourism-down-70-as-travel-restrictions-impact-all-regions

UNWTO. (n.d.a). *Impact assessment of the COVID-19 outbreak on international tourism.* https://www.unwto.org/impact-assessment-of-the-covid-19-outbreak-on-international-tourism

UNWTO. (n.d.b). *Tourism and COVID-19 – Unprecedented economic impacts.* https://www.unwto.org/tourism-and-covid-19-unprecedented-economic-impacts

Vichit-Vadakan, V. (2021, August 24). *Tourism struggles to see its future in Southeast Asia.* Heinrich-Böll-Stiftung. https://th.boell.org/en/2021/08/24/tourism-covid-19

Walters, G., Huck, L., Robinson, R. N., & Stettler, J. (2021). Commercial hospitality in tourism: A global comparison of what culturally matters. *International Journal of Hospitality Management*, *95*. https://doi.org/10.1016/j.ijhm.2021.102939

Wharton, A. (2006). Commodifying Space: Hotels and Pork Bellies. In P. Lynch, A. Morrison, & C. Lashley (Eds.), *Hospitality: A Social Lens* (pp. 101–116). Routledge. https://doi.org/10.4324/9780080465692

Wibowo, S., Widiastuti, A., & Rosardi, R. G. (2021). The role of village tourism in environmental education and social disaster management. *IOP Conference Series: Earth and Environmental Science*, *884*(1). https://doi.org/10.1088/1755-1315/884/1/012052

World Travel & Tourism Council. (2023). *Economic impact.* https://wttc.org/research/economic-impact

Yılmaz, Y., & Bititci, U. S. (2006). Performance measurement in tourism: A value chain model. *International Journal of Contemporary Hospitality Management*, *18*(4), 341–349. https://doi.org/10.1108/09596110610665348

Chapter 2

Unlocking Tourism's Potential for Business Sustainability

Winston Conrad B. Padojinog, Viory Yvonne T. Janeo and Diana Rueda

University of Asia and the Pacific, Philippines

Abstract

The tourism industry's deep interindustry linkages makes it a significant engine of growth for an economy. When aligned with specific sustainability goals, it becomes an effective means of inclusive prosperity and environmental protection. Coming fresh from the lockdown, investors and stakeholders are more conscious of their investing and spending decisions preferring more and more businesses and industries with sustainable business practices (Leal Filho et al., 2022). Specifically, tourism's generation of business and customer values must also extend to generation of value to the environment, society, and even on governance (GIZ, 2020). This chapter, using a system's view approach, demonstrates how activities in the tourism value chain – besides being an engine of growth – can also be the vehicle to attaining environmental, social, and governance (ESG) aspirations.

Keywords: Sustainability; sustainable tourism; ESG; tourism linkages; 5As of tourism

Introduction

This chapter offers a system's view of the tourism value chain which is further enriched by incorporating sustainability at each stage of the value chain. The environmental, social, and governance (ESG) is used as a framework for promoting sustainable tourism. This approach can demonstrate how value is generated and distributed within tourism's immediate environment. These impacts are illustrated using the popular Porter's 5A's of tourism: attraction, accessibility, accommodation, amenities, and activities.

Revisiting Sustainable Tourism in the Philippines, 31–43

doi:10.1108/978-1-83753-678-820241002

The first part of the chapter demonstrates the importance tourism plays in any economy and nation and how Philippine jurisprudence recognizes it. It lays down the groundwork for pushing sustainable tourism. The second part demonstrates how a system's view of sustainable tourism can demonstrate the impact tourism has on many sectors at any stage of the customer journey and experience. Properly aligned with ESG principles, sustainable tourism can help achieve sustainable development goals among the various activities in its value chain. The last part elaborates the industry's interlinkages specifically how it can be capitalized on to contribute to sustainable tourism through the ESG framework as well to meet the United Nations Sustainable Development commitments of the country in the areas of economics, ecology, and society.

Sustainable Tourism: The Philippines' Experience

The tourism industry has deep interindustry linkages established by its extensive value chain, creating a multiplier effect that has a huge impact on the country's output, employment generation, and poverty alleviation, among others (Rasool et al., 2021). In addition to these, the industry's contribution to foreign exchange earnings and foreign direct investments reflects its potential to be one of the powerful economic growth engines of a country (Tabash et al., 2023). Moreover, there are other significant and positive impacts besides economics that tourism can spell for the Philippines. And it is through these extensive linkages that tourism can be an important vehicle for contributing to the sustainability agenda (Tseng et al., 2019).

Coming out fresh from the lockdown, stakeholders, especially investors and consumers, are more militant than ever expecting sustainable business practices from the companies and the sectors they invest their funds in and purchase their goods and services from, respectively (Leal Filho et al., 2022). The generation of both business and customer values from economic activities must also extend to value generation in the environment, society, and even to governance. Tourism is one of these sectors (GIZ, 2020).

In 2022, the tourism industry accounted for 7.6% of the global GDP, creating about 22-M new jobs (World Travel & Tourism Council, n.d.). On a national scale, the share of the entire tourism industry to Philippines' total GDP showed to be remarkable at 6.2%, amounting to P1.38-T in Tourism Direct Gross Value Added (TDGVA). The 6.2% share reflected a 100 basis points increase from the previous year as tourism-related services and activities bounced back amid the consumers' revenge spending (Philippine Statistics Authority, 2023b).

Moreover, the Philippine tourism employed a total of 5.35 M (representing the 11.4% of the total employed) in 2022. It is, likewise, one of the sources of foreign exchange through the sustained growth in international tourist arrivals injecting P368.67-B to the national economy (Philippine Statistics Authority, 2023a).

The PH tourism industry, likewise, attracted 2.03-M foreign tourists in 2022 but lags far behind its ASEAN counterparts (i.e., Singapore's 6.31-M (Statista, 2023b), Thailand's 11.15-M (Statista, 2023a), and even Vietnam's 3.66-M

(Statista, 2024) in the same period). However, an overly focus on foreign tourism arrivals alone can overlook the impact that domestic tourists going around the country's different tourist destinations contribute to the economy.

By now, firms and industries engaged in tourism must be aware of their accountability for the environment, the community, and the governance footprints their tourism activities leave behind. Such so-called footprints have direct implications for the present and future generations' well-being. In fact, tourism activities must promote positive impacts rather than just preserve the status quo and must go beyond just economics (Baloch et al., 2023).

Sustainable tourism calls for a broader and deeper impact on its stakeholders and operating environment. Besides preserving and protecting the natural environment and local cultures, sustainable tourism's impact should also promote among others responsible consumption and production, reduce and eliminate income inequality, promote economic growth, and provide employment opportunities (Comerio & Strozzi, 2019).

Under Goal 8 of the UN Sustainable Development Goals (UN SDG), sustainable tourism at the minimum must have three basic tenets: ecological, economic, and sociocultural dimensions (United Nations Department of Economics and Social Affairs Division, n.d.). Citing the document *The Future We Want* from the Rio+20 outcome, these tenets were further construed and fortified. The ecological dimension includes the conservation of the environment, especially the reduction of dangerous emissions generated by the industry. Meanwhile, the economic dimension includes economic growth and infrastructure investments, and the sociocultural dimension includes the preservation and promotion of local cultural heritage (United Nations Conference on Sustainable Development, 2012).

The awareness of the role of tourism in sustainability has long been in place. Recently, the Philippine Development Plan 2023–2028 adopts WTO's definition of sustainable tourism as "tourism that takes full account of its current and future economic, social and environmental impacts, addressing the needs of visitors, the industry, the environment and host communities" (National Economic Development Authority, 2023, p. 419).

Tourism is among the sectors that can spell an extensive and deep positive impact in terms of achieving the nation's broader sustainability agenda. Its reach throughout the Philippine archipelago is extensive, and the industries it depends on to deliver its services or that benefits from its activities are strongly intertwined. Hence, tourism's sustainability practices or activities have direct implications on tourists and on the industry's vertical activities (i.e., backward and forward linkages) or supply chain. Sustainable tourism can pursue and promote greening its supply chain activities, aims for inclusive growth within its ecosystem and ensure cultural preservation and promotion.

Systems View of Tourism Business Sustainability

To fully appreciate the contribution of sustainable tourism to the national sustainable agenda, it is worthwhile to view tourism as integral and core to its immediate environment – a core with both a centripetal (i.e., inward impact) and centrifugal (i.e., outward impact) role within a system. It generates or creates value to parties and sectors within its immediate environment.

Creating a systems view of tourism business sustainability amplifies a core in value creation or generation in its value chain of industries. Hence, the tourism industry's focus on the total experience of the tourists invariably linked with the sustainable tourism agenda related to ESG becomes a primary driver of value on the different actors and industries within the system. Value generation in the sustainability context does not pertain purely on economic or financial terms only but also extends to the circular economy and inclusivity of growth; the environment in terms of the preservation and promotion, responsible consumption of goods, reusing, reducing and recycling of wastes, reduction of emissions, etc., the society in terms of cultural and heritage preservation and promotion, etc., and governance in terms of the promotion of fair competition, eradication of corruption, the promotion of human rights., etc.

Industries within the system are important at any stage of the customer journey. In this system, the industries are not seen isolated in terms of their activities and objectives in each stage of the customer journey but contribute to inclusivity where all players and stakeholders get to participate in and benefit from sustainable tourism. Competition does not become a hindrance but rather a means for these sectors to contribute to the national agenda as various establishments capitalize on the benefits that the sustainability agenda provides. Aside from the fact that sustainable practices help conserve infrastructures and destinations and help maintain these to last longer, establishments and industries committed to sustainable practices can positively impact economic, ecological and sociocultural dimensions (Moliner-Tena et al., 2020).

The impact of the tourism's essential value chain is best appreciated via *5A* (tourist's) interaction with the customers' (i.e., tourists') journey (Dickman, 1997): attraction, accommodation, accessibility, activities, and amenities. This customer journey – a stage-by-stage step tourists go through from planning to experiencing – generates not only economic value but also represents contact points that sustainable activities and practices can be promoted. After all, these 5As of tourism have one goal – that is to imprint memorable experiences to customers and "to footprint" positive impacts on the environment and society.

Tourist attraction is the ability to pull or attract travelers to a tourist destination. Such is experienced when travelers decide to leave their home to visit a destination. Citing MacCannel, tourist attraction is legitimate if three components are present: a tourist, a site to be viewed, and a marker or image which makes the site significant (Lew, 1987).

Accessibility is associated with the reachability of the destination. Various literature defines destination accessibility as how conveniently accessible a destination is. Accessibility likewise pertains to the reachability of amenities within the destination, as well as the time and effort to travel to such. Thus, such greatly affects travel decisions of tourists (Yen et al., 2021).

Accommodation is a place within the vicinity of the destination where tourists arrange to stay. It is a fundamental component which accounts for one-third of the total expenditure of the trip and is important in creating a tourism experience (Sharpley, 2000). The 2020 Survey of Tourism Establishments in the Philippines-Preliminary Results lists down the different types of accommodations which include hotels and motels, resort hotels, condotels, pension houses, camping sites or facilities, and other short-term or long-term accommodations (Philippine Statistics Authority, 2023a).

Amenities are the facilities in the tourist destination that create the travel experience. These are infrastructures such as transportation, accommodation, eating and drinking facilities and other infrastructures that are necessary to create a leisurely experience for the tourists.

Activities are unique and fun recreations which tourists look forward to in a destination. Such is the main attraction of any destination. It could range from activities (where tourists book for different activities to be done in the destination) or simply sightseeing in the destination. Such is the highlight of the customer journey.

This push-and-pull concept in tourism is centered on the idea of enjoyment people find as they travel. Mitas and Bastiaansen (2018) suggest that enjoyment is in the novel experience – something new or different from their usual day-to-day activities. The tourism industry itself and its linkages connected to the customer journey – information and infrastructure, hotels and lodgings, transportation, tour operators and guides and natural environment, cultural activities, and facilities – all play a crucial role.

The system's view of such an impact is influenced by the national agenda. In the Philippines, the tourism law (RA 9593) as well as Goal 8.9 of the Sustainable Development Goals of the United Nations highlighted the following:

> We encourage the promotion of investment in sustainable tourism, including eco-tourism and cultural tourism, which may include creating small and medium-sized enterprises and facilitating access to finance, including through microcredit initiatives for the poor, indigenous peoples and local communities in areas with high eco-tourism potential. In this regard, we underline the importance of establishing, where necessary, appropriate guidelines and regulations in accordance with national priorities and legislation for promoting and supporting sustainable tourism. (Conference on Sustainable Development, 2012, par. 131)

Moreover, the Philippine Development Plan 2023–2028 further accelerated and intensified campaign for sustainable tourism by promoting "sustainable, inclusive, and resilient multidimensional tourism and holiday economics" (National Economic Development Authority, 2023, p. 170). Such a plan includes highlighting the strengths of Filipino services, culture and heritage, arts, food, gastronomy (including Halal), and nature-based tourism, among others. These plans include collaborations with other government agencies and Local Government Units (LGUs).

The UN World Tourism Organization (UNWTO), on the other hand, defines sustainable tourism as "tourism that takes full account of its current and future economic, social and environmental impacts while addressing the needs of visitors, the industry, the environment and host communities" (UNWTO, 2023, p. 6).

Thus, sustainability further elaborates the responsible utilization of environmental resources which includes its conservation, the promotion and conservation of heritage and the promotion of long-term economic growth that will provide socioeconomic benefits to all stakeholders (UNWTO, 2023).

Tourism Linkages and Its Role in Sustainable Tourism

The so-called greening the tourism value chain and its activities, especially in its linkages or supply chain, can accelerate meeting or even promoting the ESG objectives. UNWTO further elaborated such a concept as it includes: (1) optimal use of natural resources which includes maintenance and preservation of ecological processes, natural resources, and biodiversity. This is demonstrated as sustainable tourism generates value and further accelerate; (2) giving value to the sociocultural authenticity of host communities through promotion and preservation of traditional values and living cultural heritages; and (3) securing a sustainable long term economic benefits to all stakeholders manifested by fairly distributed growth (UNWTO, 2023). Thus, sustainable tourism generates incentives and greater value in economic, ecological, and sociocultural dimensions.

Some of the few Philippine examples are highlighted below:

Economic Dimension. Using the latest 80 × 80 Leontief Inverse Matrix Input Output Table published by PSA in 2021 and using the Food and Beverage (F&B) service sector as a proxy to represent the tourism sector (i.e., the sector accounts as the top tourism-related expenditure and the largest subindustry of tourism) (Philippine Statistics Authority, 2023a), the economic dimension of sustainable tourism is best appreciated in terms of the strengths of its backward and forward linkages as well magnitude of its total output. Such forward and backward linkages in the supply chain play a key role in generating the multiplier impacts. Table 2.1 summarizes the top 10 input (i.e., backward linkages) and output (i.e., forward linkages) sectors of the F&B services, respectively. The forward and backward linkages and output multipliers were computed using the approach of Blair and Miller (2009).

Except for accommodations, arts and entertainment, and nonbanks, all of the sectors in Table 2.1 demonstrate above average backward linkages to F&B services (refer to fourth column). Note that indices of less than one indicate weak backward linkage while indices greater than one indicate strong backward linkages. Relating these above sectors vis-a-vis with the *5A's* of tourism – attraction, accommodation, accessibility, amenities, and activities – attests to the ability of tourism activities to generate economic value all throughout the tourists' journey. This strong backward linkage exhibits the potential impact industries related to tourism (i.e., hotels, accommodations, information and infrastructures, etc.) can generate for the economy.

Table 2.1. Top 10 Inputs and Outputs Using Food and Beverage Service Activities as Proxy for Tourism and Backward Linkages and Selected Sector Output Multipliers.

Inputs Using Food and Beverage Service Activities as Proxy for Tourism and Backward Linkages and Selected Sector Output Multipliers								Outputs Using Food and Beverage Service Activities as Proxy for Tourism and Forward Linkages and Selected Sector Output Multipliers							
Industry	**Inputs**	**Ranking Based on Input**	**Total BL**	**Ranking Based on BL**	**Total FL**	**Ranking Based on FL**	**Output Multiplier**	**Industry**	**Outputs**	**Ranking Based on Output**	**Total BL**	**Ranking Based on BL**	**Total FL**	**Ranking Based on FL**	**Output Multiplier**
Food and beverage service activities	1.1803	1	1.2340	16	1.2663	44	2.4003	Food and beverage service activities	1.1803	1	1.2340	16	0.6508	44	2.4003
Accommodation	0.0227	2	0.9566	44	1.0334	74	1.8607	Retail trade, except of motor vehicles and motorcycles	0.2579	2	0.7891	62	3.4170	3	1.5349
Arts, entertainment and recreation	0.0065	3	0.7056	70	1.1765	54	1.3725	Food products	0.2318	3	1.1748	21	2.4837	6	2.2851
Nonbanks	0.0065	4	0.8336	57	2.2735	17	1.6214	Chemical and chemical products	0.0962	4	1.2996	12	3.5279	2	2.5278
Social work activities	0.0038	5	1.0137	41	1.0199	76	1.9718	Professional, scientific and technical activities; administrative and support service activities	0.0633	5	0.9132	51	2.7458	4	1.7762

(Continued)

Table 2.1. *(Continued)*

Inputs Using Food and Beverage Service Activities as Proxy for Tourism and Backward Linkages and Selected Sector Output Multipliers								Outputs Using Food and Beverage Service Activities as Proxy for Tourism and Forward Linkages and Selected Sector Output Multipliers							
Industry	**Inputs**	**Ranking Based on Input**	**Total BL**	**Ranking Based on BL**	**Total FL**	**Ranking Based on FL**	**Output Multiplier**	**Industry**	**Outputs**	**Ranking Based on Output**	**Total BL**	**Ranking Based on BL**	**Total FL**	**Ranking Based on FL**	**Output Multiplier**
Land transport	0.0019	6	1.1079	27	2.8194	13	2.1549	Wholesale trade, except of motor vehicles and motorcycles	0.0599	6	0.7763	63	1.9870	9	1.5100
Water transport	0.0018	7	1.0855	32	1.2229	50	2.1114	Real estate	0.0587	7	0.8113	60	2.7084	5	1.5781
Air transport	0.0017	8	1.3894	3	1.2229	50	2.7026	Coke and refined petroleum products	0.0525	8	1.2113	18	4.1260	1	2.3560
Transport equipment	0.0015	9	1.3309	8	1.2368	48	2.5887	Electricity	0.0330	9	0.9214	50	1.7977	10	1.7923
Communication	0.0014	10	1.0164	40	2.2558	18	1.9770	Computer, electronic and optical products	0.0306	10	1.3078	11	1.5612	12	2.5438

Source: Computed and derived from the 2021 PSA Input-Output Matrix.

Note: BL (backward linkages), FL (forward linkages).

Table 2.1 reflects the forward linkages of the F&B sector as well. Just like in the backward linkages analysis, the indices of less than one indicate a weak forward linkage and vice versa. Except for the F&B service sector, all linkages with F&B have strong forward linkages such as retail trade, food products, chemical products, etc. (refer to 6th column).

The F&B sector's output multiplier of 2.4 (see Table 2.1, 8th column) implies that a 1.0 peso increase in spending in the sector will trigger 1.4 pesos increase in the output of the other related sectors or 2.4 pesos increase for the rest of the economy (i.e., including the 1.0 peso spent on F&B services sector). Among these sectors, those related to transportations have the highest multipliers. In terms of the 5A's, the major modes of transport have significant impacts – air transport and transport equipment – at 2.7 and 2.6 respectively (see Table 2.1, 8th column). Consequently, this implies that a 1.0 peso increase in spending in these sectors increases by 1.7 pesos and 1.6 pesos respectively the outputs of other related industries or 2.7 pesos and 2.6 pesos the total economic output. These sectors are actively engaging with many stakeholders who can spell inclusive growth.

Meanwhile, nonbanking financial institution (NBFI), which provides SMEs and *un-banked individuals* catering to tourism, access to alternative financial services such as investments, brokering, money remittances, and loans – all promoting financial inclusion in the countryside. Meanwhile, online tourism services and search engines that promote travel destinations to a wider audience not only better the customer experience but also generate savings and reduce footprints at the fraction of the cost (Koliouska & Andreopoulou, 2023).

Ecological Dimension. Mindful engineering adopted by new projects can have significant environmental impact such as the case of the 400-hectare Masungi Georeserve in Baras, Rizal. The DENR and a private corporation (Blue Star Construction & Development Corporation) entered into a joint venture for a conservation-oriented development project in 1996 when the area was barren due to the presence of informal settlers and illegal loggers. Since then, the "georeserve" has transformed with the natural forest restored such that it has become an important tourist destination (Dumaliang et al., 2018). Thanks to mindful engineering coupled with strict policy enforcement, careful planning of the structural designs and the materials used and the installations of the temporary structures inside the conservation area (Dumaliang, 2020) the sanctuary's original state was restored and now preserved.

Another example is the adoption of e-wallets by stakeholders in tourism driven by fintech companies. This did not only facilitate financial inclusivity but also to the reduction of environmental pollution through *cashless, cardless and paperless* transactions (Hopali et al., 2022). The lockdown contributed to the accelerated digital transformation of various industries including tourism (Armutcu et al., 2023) giving rise to e-tourism – the use of Information and Communication Technologies (ICT) in the tourism industry (Hopali et al., 2022). Internet and other online digital technologies promote, market and sell tourism-related products and services, offering a wide array of choices of accommodation (Hopali et al., 2022) and transportation (Ganzon, 2021). The same online platforms become avenues for information and education campaign for communicating sustainable tourism practices, reducing the carbon footprints by providing virtual tours, augmented reality experiences and interactive digital content (Susilo & Santos, 2022).

Stakeholder involvement in sustainable tourism promotes synergies (Heslinga et al., 2019, as cited in Roxas et al., 2020) which Persha et al. (2011, as cited in Roxas et al., 2020) pointed out to have facilitated the interactions among them to achieve societal and ecological goals. This collaboration spawned environmental protection and tourism development as complementary and not as competing outcomes (Roxas et al., 2020).

Sociocultural Dimension. Stakeholder engagement is also essential in the sociocultural dimension of sustainable tourism. The national government plays an important role in facilitating such engagements and must encourage tourism investments by investing in public goods such as roads and bridges, peace and order, stable and clear policies and promoting investments in physical and digital networks, and affordable utilities, among others (Amaoko et al., 2022).

The stakeholder community is composed of different households and individuals – "smallest units of decision-making and activity – are frequently overlooked in the engagement process (Uphoff, 1992, p. 3)." The impact and their expectations must be taken into account and met; their welfare and just benefits must be protected and ensured. Thus, community participation is an essential party in any stakeholders' engagement in order to achieve sustainable development (Gutierrez, 2019).

For instance, Boracay Island, globally known for its white sandy beaches and dependence on tourism, suffered from *over-tourism*. Waste and other environmental pollution due to the lack of sewer and solid waste infrastructures and violation of land regulation laws threatened the ecological balance and thus the attractiveness of the sight as a tourist destination (Reyes et al., 2018). After the 2018 government-enforced closure of the island, the national and local government together with the private sector and local communities collaborated to implement projects and enforced better policies and practices to make the island once again sustainable and attractive for tourism. Since opening in September 2021, the island is now on more solid footing while growing its tourism base in conjunction with maintaining environmental harmony and community well-being.

The sociocultural aspect of sustainable tourism can also bring about good governance. This has been evident in several cases in popular tourism sites. Again, citing the Boracay rehabilitation program, a stronger commitment for sustainable ecotourism practices to rehabilitate and conserve Boracay Island pushed the LGUs and the National Government (NG) to be more proactive in reducing corruption, bureaucracy, and ensure proper implementation of the policies that preserves the natural beauty of the island, enforce laws to encourage visitors and establishments to adhere to the local policies and ordinances. The succeeding years were spent on the rehabilitation of the island: plans were formulated and implemented, fundings were allocated to the appropriate initiatives, and continuous coordination between and among the LGU, and the Department of Tourism (DOT) and the Department of Environment and Natural Resources (DENR), reduction of corruption and the issuance of permits.

Conclusion

Providing the best customer experience to both foreign and domestic tourists can become a key driver in promoting the sustainable agenda and practices through sustainable tourism. Because of tourism's extensive and deep industry linkages and extensive network of stakeholders, the tourists' customer journey should be an effective vehicle to promote sustainable tourism. This further establishes how sustainable tourism is grounded on and related with the ESG principles and practices or activities in the value chain.

It offers a fresh perspective on business sustainability related to tourism introducing a systems view perspective that highlights how sustainable tourism, through ESG principles, can generate systemic economic value and bring about environmental conservation and promotion, grow and support local and national economies, and preserve and promote cultural heritage and good governance.

Thus, it is important to capitalize on the opportunities that sustainable tourism offers properly guided and encouraged by a clear and stable policy environment, with individuals and establishments inspired by the sustainable agenda, by an informed, empowered, and responsible travel behavior of visitors and engaged stakeholders.

References

Amaoko, G. K., Obuobisa-Darko, T., & Marfo, S. O. (2022). Stakeholder role in tourism sustainability: The case of Kwame Nkrumah Mausoleum and centre for art and culture in Ghana. *International Hospitality Review*, *36*(1), 25–44. https://doi.org/10.1108/IHR-09-2020-0057

Armutcu, B., Tan, A., Amponsah, M., Parida, S., & Ramkissoon, H. (2023). Tourist behaviour: The role of digital marketing and social media. *Acta Psychologica*, *240*. https://doi.org/10.1016/j.actpsy.2023.104025

Baloch, Q. B., Shah, S. N., Iqbal, N., Sheeraz, M., Asadullah, M., Mahar, S., & Khan, A. U. (2023). Impact of tourism development upon environmental sustainability: A suggested framework for sustainable ecotourism. *Environmental Science and Pollution Research*, *30*, 5917–5930. https://doi.org/10.1007/s11356-022-22496-w

Blair, R., & Miller, P. (2009). *Input-output analysis foundations and extensions* (2nd ed.). Cambridge University Press.

Comerio, N., & Strozzi, F. (2019). Tourism and its economic impact: A literature review using bibliometric tools. *Tourism Economics*, *25*(1), 109–131. https://doi.org/10.1177/1354816618793762

Deutsche Gesellschaft für Internationale Zusammenarbeit (GIZ) GmbH. (2020). *The tourism value chain*. https://www.switch-asia.eu/site/assets/files/2460/giz_tourism_value_chains_en.pdf

Dickman, S. (1997). *Tourism: An introductory text*. Hoddey Headline.

Dumaliang, B. (2020). The georeserve model: Mindful engineering as a means to build sustainable tourism value and income in conservation areas. *Panorama: Solutions for a Healthy Planet*. https://panorama.solutions/en/solution/georeserve-model-mindful-engineering-means-build-sustainable-tourism-value-and-income

Dumaliang, A., Dumaliang, B. C., & Pena, R. (2018). Philippines: The case of Masungi georeserve. In R. Dowling & D. Newsome (Eds.), *Handbook of geotourism* (pp. 448–461). Service Filmsetting Ltd.

Ganzon, M. K. (2021). Discovering the role of informal transport in tourism in developing countries. *Journal of the Eastern Society for Transportation Studies, 14*. https://doi.org/10.11175/easts.14.2470

Gutierrez, E. L. (2019). Participation in tourism: Cases on community-based tourism (CBT) in the Philippines. *Ritsumeikan Journal of Asia Pacific Studies, 37*. https://doi.org/10.34409/rjaps.37.1_23

Hopali, E., Vayvay, Ö., Kalender, Z. T., Turhan, D., & Aysuna, C. (2022). How do mobile wallets improve sustainability in payment services? A comprehensive literature review. *Sustainability, 14*(24). https://doi.org/10.3390/su142416541

Koliouska, C., & Andreopoulou, Z. (2023). E-Tourism for sustainable development through alternative tourism activities. *Sustainability, 15*, 8485. https://doi.org/10.3390/su15118485

Leal Filho, W., Salvia, A. L., Paço, A., Dinis, M. A. P., Vidal, D. G., Da Cunha, D. A., de Vasconcelos, C. R., Baumgartner, R. J., Rampasso, I., Anholon, R., Doni, F., Sonetti, G., Azeiteiro, U., Carvalho, S., & Rios, F. J. M. (2022). The influences of the COVID-19 pandemic on sustainable consumption: An international study. *Environmental Sciences Europe, 34*(54). https://doi.org/10.1186/s12302-022-00626-y

Lew, A. (1987). A framework of tourist attraction research. *Annals of Tourism Research, 14*, 553–575. https://doi.org/10.1016/0160-7383(87)90071-5

Mitas, O., & Bastiaansen, M. (2018). Novelty: A mechanism of tourists' enjoyment. *Annals of Tourism Research, 72*, 98–108. https://doi.org/10.1016/j.annals.2018.07.002

Moliner-Tena, M. Á, Monferrer-Tirado, D., Ferreres-Bonfill, J. B., & Rodríguez-Artola, R. M. (2020). Destination sustainability and memorable tourism experiences. *Sustainability, 13*(21), 11996. https://doi.org/10.3390/su132111996

National Economic Development Authority. (2023). *Philippine Development Plan 2023–2028*. https://pdp.neda.gov.ph/wp-content/uploads/2023/01/PDP-2023-2028.pdf

Philippine Statistics Authority. (2021, December 9). *2019 80 x 80 input output table*. https://psa.gov.ph/statistics/supply-and-use-input-output

Philippine Statistics Authority. (2023a, January 18). *2020 Survey of Tourism Establishments in the Philippines (STEP) – Economy-wide: Preliminary results*. https://psa.gov.ph/statistics/step

Philippine Statistics Authority. (2023b, June 15). *Tourism Contributes 6.2 Percent to GDP in 2022*. https://psa.gov.ph/system/files/sad/ons-cleared_1.-revised_1-PRESS-RELEASE_2022-PTSA-16-June_ONSF2-signed_0.pdf

Rasool, H., Maqbool, S., & Tarique, M. (2021). The relationship between tourism and economic growth among BRICS countries: A panel cointegration analysis. *Future Business Journal, 7*(1). https://doi.org/10.1186/s43093-020-00048-3

Reyes, C., Albert, J. R., Quimba, F. M., Ortiz, M. K., & Asis, R. (2018). *The Boracay closure: Socioeconomic consequence and resilience management*. Philippine Institute of Development Studies Discussion Paper Series 2018-37. https://pidswebs.pids.gov.ph/CDN/PUBLICATIONS/pidsdps1837.pdf

Roxas, F. M., Rivera, J. P., & Gutierrez, E. L. (2020). Mapping stakeholders' roles in governing sustainable tourism destinations. *Journal of Hospitality and Tourism Management, 45*, 387–398. https://doi.org/10.1016/j.jhtm.2020.09.005

Sharpley, S. (2000). The influence of the accommodation sector on tourism development: Lessons from Cyprus. *International Journal of Hospitality Management*, *12*(30), 275–293. https://doi.org/10.1016/S0278-4319(00)00021-9

Statista. (2023a, September 26). *Number of international visitor arrivals in Thailand from 2015 to 2022*. https://www.statista.com/statistics/994693/thailand-number-international-tourist-arrivals/

Statista. (2023b, November 22). *Number of international visitor arrivals in Singapore from 2013 to 2022*. https://www.statista.com/statistics/977993/total-international-visitor-arrivals-singapore/

Statista. (2024, February 8). *Number of international visitor arrivals in Vietnam from 2010 to 2023*. https://www.statista.com/statistics/981987/vietnam-number-international-tourist-arrivals/

Susilo, D., & Santos, M. C. K. (2022). Digital marketing communication for promoting Philippine tourism. *International Journal of Research and Review*, *10*(6). https://doi.org/10.52403/ijrr.20230625

Tabash, M., Anagreh, S., Subhani, B. H., Al-Feryan, M. A. S., & Drachal, K. (2023). Tourism, remittances, and foreign investment as determinants of economic growth: Empirical evidence from selected Asian economies. *Economies*, *11*(2), 54. https://doi.org/10.3390/economies11020054

Tseng, M. L., Tan, P. A., Jeng, S. Y., Lin, C. W. R., Negash, Y. T., & Darsono, S. N. A. C. (2019). Sustainable investment: Interrelated among corporate governance, economic performance and market risks using investor preference approach. *Sustainability*, *11*(7), 2108. https://doi.org/10.3390/su11072108

United Nations Conference on Sustainable Development. (2012, June). *Future we want – Outcome document*. https://sustainabledevelopment.un.org/futurewewant.html

United Nations Department of Economics and Social Affairs Division. (n.d.). *Sustainable tourism*. https://sdgs.un.org/goals/goal8

United Nations World Tourism Organization. (2023, October). *Statistical Framework for Measuring the Sustainability of Tourism (SF-MST)*. https://webunwto.s3.eu-west-1.amazonaws.com/s3fs-public/2023-10/SF-MST_complete_version_OCT2023_cover03.pdf

Uphoff, N. (1992, January 1). *Local institutions and participation for sustainable development*. International Institute for Environment and Development. http://www.jstor.org/stable/resrep01670

World Travel and Tourism Council. (n.d.). *Economic impact research*. https://wttc.org/research/economic-impact#:~:text=In%202022%2C%20the%20Travel%20%26%20Tourism,%2C%20only%2014.1%25%20below%202019

Yen, H. P., Chen, P. C., & Ho, K. C. (2021). Analyzing destination accessibility from the perspective of efficiency among tourism origin countries. *Sage Open*, *11*(2). https://doi.org/10.1177/21582440211005752

Chapter 3

The Tourism Value Chain as Framework for Sustainable Tourism Economies: Challenges and Future Opportunities for Destinations

Ramon Benedicto A. Alampay and Omme Atiyah B. Gonting

University of the Philippines Diliman, Philippines

Abstract

Tourism value chain (TVC) analysis has been widely used as an alternative approach for designing sustainable and inclusive programs for tourism development in Africa, Asia, and the Pacific. This chapter looks at the experiences of three sustainable tourism initiatives in the Philippines to assess the strengths and limitations of the value chain as a framework for destination development. Short case studies describe the experiences of stakeholders in two popular destinations in the Philippines: the surfing town of San Juan, La Union, and the resort-island of Panglao, Bohol. The third case study shares the perspective of the Transforming the Tourism Value Chain (TTVC) project, a national campaign for more sustainable hotels, resort as well as meetings, conventions, incentives, and event (MICE) facilities in various destinations around the country. Stakeholders deeply involved in implementing these projects share their insights on the successes, challenges, and limitations of TVC-based approaches to resilient and sustainable destination development. The destinations' experiences suggest that the advocacy for sustainable management and operations may be comparable to a process of technology adoption or acceptance along the TVC. Practical and theoretical recommendations for leveraging the TVC toward more holistic and sustainable visitor economies are given at the end of this chapter.

Keywords: Tourism value chain; sustainable tourism; destination development; Philippines; value chain analysis

Revisiting Sustainable Tourism in the Philippines, 45–65

doi:10.1108/978-1-83753-678-820241003

Background

As global travel and tourism steadily re-emerged from the COVID-19 pandemic, it finds itself at a crossroads where one fork turns back toward a prepandemic condition characterized by an unsustainable growth-oriented trajectory (Ioannides & Gyimóthy, 2020). The emphasis on growth, even when embedded in the paradigm of sustainable development, had resulted in a consumption-based, industrialized system that resulted in degenerative economies, damaged natural environments, and an ever-increasing gap between the haves and the have-nots of the world. Thus, there have been increasing and louder calls – even before the pandemic – that the global economic order was seriously flawed and in need of radical change (Du Plessis, 2017; Goodwin, 2017; Raworth, 2017b).

In many destinations, rampant tourism growth was been met with hostility and outright violence, as the quality of life in the area – as well as the quality of the visitor experience – deteriorated from too many visitors (Goodwin, 2017). Overtourism led to calls that the anthropogenic model of economic growth represented by sustainable tourism needed to be replaced by more resilient, regenerative alternatives (Cave & Dredge, 2018; Du Plessis, 2017; Raworth, 2017a; Sharpley, 2020; Sheldon, 2022).

Ironically, COVID-19 had given tourism a much-needed opportunity to pause and consider the alternative path that radically departs from the usual practices and excesses of the so-called prepandemic Normal (Ioannides & Gyimóthy, 2020). We look at the opportunities for doing so from the perspective of tourism as the mode by which a destination's product, i.e., visitor experiences, is produced (Pearce, 2014). This production process requires a system of interactions between tourists, the enterprises directly and indirectly involved in the delivery of tourist services, as well as the residents of the destinations themselves (Jovicic, 2016). This chapter applies a value chain perspective to this process in examining how Asian tourism industries might transition to the transformative, circular visitor economies that have been envisioned for the so-called Next Normal of tourism. Using examples of tourism value chain (TVC)-based programs and interventions in select Philippine destinations, it draws lessons for advancing more sustainable forms of tourism development moving forward.

The Production of Visitor Experiences Tourism as a Value Chain

The United Nations World Tourism Organization (UNWTO) defines the TVC as "a sequence of primary and support activities that are strategically fundamental to the performance of the tourism sector" (World Tourism Organization, 2018, p. 31). Originally proposed by Porter (1985 as cited in Song et al., 2013) as a microlevel tool for identifying a firm's competitive advantages (Romero & Tejada, 2011), it has since been extended to a macro-level analysis of the relationships among the different actors performing the value-adding activities needed to bring an industry's product or service to market (Song et al., 2013). In tourism, it has become a fairly common framework for analyzing the development potentials and constraints of local tourism industries around the world (Pilarca, 2008).

Fig. 3.1 shows a typical depiction of a TVC for a given destination. The resulting product represents a portfolio of inter-related services that a tourist requires when traveling to and staying in the destination. The value-adding process follows the tourist points of contact with service providers throughout their customer journey (Partale, 2020). This starts with their trip planning and preparation activities and continues to their actual activities at the destination, from arrival to departure. Government and nongovernmental influence groups regulate and interact with the service providers throughout the process' various stages.

Partale (2020) points out that each link in the chain is equivalent to a particular subsector or type of tourism service. In turn, each of these subsectors, e.g., accommodations, food and beverage, etc., requires its supply chain of goods and services to deliver the expected service or experience. Thus, the tourism production process is more akin to a complex value creation system than the simplistic, linear process that the imagery of a "value chain" might imply.

TVC analysis has also been used to develop a more holistic evaluation of the destination and tourism's contribution to sustainable development, especially in emerging economies (Partale, 2020). It has been acknowledged as a useful tool for identifying opportunities for improving and enhancing the participation of disadvantaged groups (Hummel et al., 2013; World Tourism Organization, 2018) and for looking at environmental and social outcomes of the TVC activities (Lara-Morales & Clarke, 2022). TVC analysis was a preferred approach by international development agencies such as the SNV Netherlands Development Organization, Overseas Development Institute (ODI), and the International Trade Centre (ITC) in exploring the extent to which poor people were incorporated into tourism in African destinations (Spenceley & Meyer, 2012).

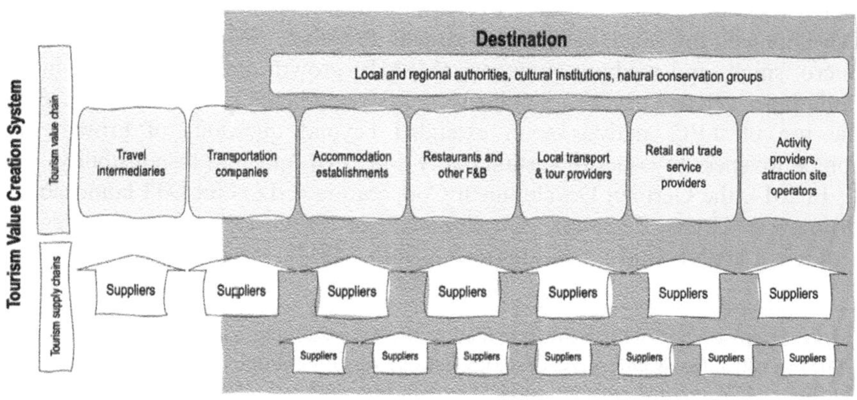

Fig. 3.1. The Tourism Value Creation System. *Note:* Author constructed based on Partale (2020).

Similarly, the Netherlands Development Organization (SNV) used a TVC analysis approach to inform their sustainable tourism development programs in designing their pro-poor tourism initiatives in Southeast (Laos and Vietnam) and South Asia (Nepal and Bhutan) from 2006 to 2010. Before this period, SNV had focused its assistance to Laos on product development, to support the government's focus on community-based ecotourism. As a result, almost all their activities were limited to rural areas. After commissioning the ODI to undertake a value chain analysis of Luang Prabang in 2006, SNV realized that the product development interventions would not be sustainable and the poverty alleviation outcomes would not be optimized without integrating destination development and management into their interventions (SNV, 2012). Following this first study, SNV adopted what it called a "VCA and Development" framework where VCA would be the starting point for their pro-poor interventions in Laos. Doing so enabled SNV to develop more holistic programs that engaged the tourism private sector beyond the rural areas of the destination region. By also developing the destination development and management capacities of regional stakeholders, SNV increased the development impact from scaling up their community-based tourism projects to more multi-province circuits or destination regions (Hummel et al., 2013; SNV, 2012). The VCA and Development approach guided subsequent SNV projects in Vietnam, Bhutan, and Nepal.

Case Studies in TVC Approaches to Sustainable Tourism Development

Value chain analysis has been applied as an evaluation and planning tool in Philippine tourism since the German Technical Cooperation (GTZ) introduced the approach in 2007 in analyzing the health-, spa-and-wellness-, education- and eco-tourism sectors in the central region of the Philippines. TVC analysis was one of the principal tools that GTZ used in codesigning interventions, jointly with the Department of Trade and Industry (DTI), for their Small and Medium Enterprise Development for Sustainable Employment Program (SMEDSEP) to promote micro, small, and medium enterprise (MSME) growth and development in the Philippines (Pilarca, 2008).[1] Similar to the developments in other parts of Asia, the use of TVC analysis soon extended beyond questions of growth and competitiveness to issues of sustainability and environmental responsibility.

In 2014, the German Development Cooperation (GIZ) and DTI launched the Promoting Green Economic Development or ProGED project to integrate ecological aspects and green approaches into MSME development.[2,3] ProGED

[1]More details on SMEDSEP are provided in the case study by Vahlhaus and Macaranas (2005) at https://www.enterprise-development.org/wp-content/uploads/Valhausdoc.pdf

[2]In 2011, GTZ merged with two other German international development organizations, the German Development Service (DED) and Capacity Building International (InWEnt), to form the unified organization that is GIZ.

[3]More details on the program are provided in the ProGED factsheet at https://tuewas-asia.org/wp-content/uploads/2017/06/Factsheet.pdf

included a specific component for the tourism sector, the Greening the Tourism Value Chain project, which was implemented principally in the province of Bohol (Antonio, 2012; Finkel et al., 2010).

In 2017, the Philippines was one of four countries included in the United Nations Environment Programme's (UNEP) Transforming Tourism Value Chains (TTVC) Project, which had the aim of reducing greenhouse gas (GHG) emissions and improving resource efficiency in the three tourism sectors with the highest usage of water and energy resources: accommodations, food and beverage (F&B), and meetings, incentives, conferences, and events (MICE). In contrast to the more destination-focused GIZ project, this was a nationwide project, although many of the participating establishments were based in the capital region of Metro Manila.

In this section, we look at the experiences and lessons learned from these two projects, which were explicitly designed around the TVC. A third case study seeks insights from the sustainable tourism development efforts for the surfing town of San Juan in La Union province. Unlike the first two cases, these efforts involved organically from local tourism groups and were not overtly based on a value chain framework. However, the San Juan experience – involving different tourism industry components (often working in concert) – provides an illustrative example of a destination value chain in action.

These cases are based on observations and reflections shared by key participants and players in each of these programs.

Greening the TVC in Panglao Island

Panglao Island is the primary resort destination of the province of Bohol. The province is located in the central region of the Visayas, a cluster of islands that cuts across the middle of the Philippine archipelago. The Visayas island group is acknowledged to be the location of the most important sun-sand-and-sea destinations in the country. These include Boracay, Cebu, Palawan, and Bohol, all of which are only an hour or 90 minutes away from Manila by air.

The 91-square-kilometer Panglao Island, just a bridgeway across from mainland Bohol, is known for its white beaches and world-class dive sites and accounts for at least 80% of all international arrivals to the province. Even after the COVID-19 pandemic, the hotel-resort sector in Panglao is still expected to expand to nearly 3,500 rooms by 2027 with new investments expected in larger, upscale properties (Aw & Phuttitarn, 2020).

However, Panglao Island's rise to becoming one of the Philippines' most popular island resort destinations placed it on the list of the national government's "tourism hotspots" at high risk of environmental decline in 2018. It was at risk of experiencing the same fate as the popular Boracay Island which the national government had closed down from March to October that year, citing problems of overdevelopment, poor solid waste and sewage management, water pollution, and other ecological problems (Ramirez, 2018). This was particularly concerning for Bohol which was aspiring to be the first UNESCO Geopark in the country (Businessworld, 2019).

Interestingly, Bohol – and specifically, Panglao Island – was one of the sites chosen by the GIZ for its Greening the Tourism Value Chain project in 2014. That was a component of a green economic development promotion program (ProGED) which sought to integrate ecological aspects and green approaches into the DTI's earlier program to strengthen MSMEs in the TVC (Antonio, 2012; Finkel et al., 2010).

It did this by working with the Bohol Association of Hotels, Resorts, and Restaurants (BAHRR) to build member hotels' capacities to institutionalize green practices along six functional areas identified through value chain analysis. These were: transportation, supply and procurement, natural resource management, energy efficiency, water conservation, and solid waste management – the last three of which were specific to hotel and resort operations (Finkel et al., 2010). The project trained about 50 BAHRR members on installing or upgrading their environmental management systems and organized events to match the hotels and resorts with providers of green services and technologies. Other capacity-building workshops focused on energy efficiency and renewable energy, green financing, and orientations on the ASEAN Green Hotel certification program (Arbon, 2014).

One of these hotels, the Amarela Resort, went on to earn the ASEAN certification from 2014 to 2016.[4] Mr Lucas Nunag, Amarela's owner and also Chair of Bohol's Provincial Tourism Council, considered the overall impact of the project to be positive in the sense that public and private stakeholders were educated about green technologies and approaches to tourism. Amarela and two other Panglao hotels, the Bellevue and the Amorita, have been able to sustain their green efforts such that they have each received the ASEAN Green Hotel award on three separate occasions. For this, all three properties were elevated to the ASEAN Hall of Fame Award for Green Hotels in 2023.

However, Mr Nunag acknowledged that the more visible and lasting benefits of the Greening the Tourism Value Chain project could be observed in individual hotels and resorts more than in the destination of Panglao as a whole. According to Mr Nunag, the hotels and resorts that participated in the project were able to continue applying what they learned to improve their efficiency in waste management and energy conservation.

> In the area of water management, we started with the reed bed system. In time, DENR upgraded their parameters and you need to remove certain amount of ammonia, potassium. So I had to upgrade to a moving bed bioreactor. [It involved] some capital expense but it was very efficient. And then solar. We have solar for energizing the whole operation during the day. We have a limited battery on standby and to power our wastewater motors during the night. Plus, we are improving and enhancing the farm to table,

---

[4]Based on 11 criteria: environmental policies, use of green products, collaboration with the community and local organizations, human resource development, solid waste management, energy efficiency, water efficiency, air quality management, noise pollution control, wastewater treatment and chemical substance disposal.

> adding more items to the garden, and coming up with more soil enhancements through vermiculture and biochar.[5] (L. Nunag, personal communication, October 12, 2023)

The Amorita has formally institutionalized its environmental program as its Project ECHO which includes integrating energy efficiency and green design into the property as well activities to build and strengthen linkages with the community by developing and employing local workforce, purchasing products and services from local suppliers, and organizing community-wide education and capacity-building activities.[6] The Bellevue Resort has also been active in spreading its environmental conservation activities beyond its property walls. It has organized various information, education, and promotion events such as the annual International Coastal Clean-Up and the Lights-Off Concert to commemorate International Earth Hour. In addition, it has partnered with the Department of Tourism (DOT) on a Native Giant Clams Rescue Project to protect and cultivate the endangered mollusk locally known as *taklobo* (Gutierrez, 2023).

For Mr Nunag, projects like Greening the Tourism Value Chain are very useful because the participating resorts are not the only ones to learn. Strong industry associations, such as BAHRR, can then facilitate effective implementation through the peer support and pressure that members give each other, as well as the demonstration effects of the pioneering participants.

> I hope similar efforts can return. Because after that, the resorts learn (and) the associations encourage their members to observe and emulate each other's good practices. (L. Nunag, personal communication, October 12, 2023)

Despite individual successes, Panglao Island remains vulnerable, particularly around its popular Alona Beach. According to Mr Nunag, some of the red flags from Boracay are noticeably present in Panglao. Long-running problems of solid waste and sewage management, overdevelopment, and beach erosion remain even as foreign visitor arrivals have not fully returned to pre-COVID-19 volumes. Foreign visitor arrivals to Bohol in 2023 were projected to reach approximately 250,000 by the end of 2023 (Arnaldo, 2023 November 28) or about 34% of the foreign arrival figures in 2019.

Foreign visitors accounted for 46% of the 1,581,904 total arrivals to the province in 2019. Two countries, China (45%) and South Korea (16%), generated roughly three out of every five foreign arrivals in that year (Source: Bohol Tourism Office). While the South Korean market for Bohol had returned to nearly 90% of

[5]Biochar is produced through a controlled burning or pyrolysis of agricultural and food waste (biomass). Biochar can then be used to enrich the soil.

[6]ECHO stands for Environment, Community and Heritage by One-Of Collection (the hotel management group that Amorita is part of). For more details, see https://www.amoritaresort.com/project-echo.php

pre-COVID-19 numbers in November 2023 (Arnaldo, 2023), Chinese arrivals to the entire Philippines remained at a paltry 15% of pre-COVID-19 volume (Source: Department of Tourism).

Although he credits the local government units (LGU) in Bohol for being serious about segregation and collection of solid waste, there are destination-wide problems of logistics and social behavior that hamper those efforts and prohibit sustained implementation. Local government efforts will only be as successful as the actions of individual residents and visitors.

> There are already some ordinances on segregation, composting of biodegradables to be composted, etc. But if you walk along the beach, you'll still see a lot of plastics and nonbiodegradables. Libaong Beach is one of the last remaining public beaches in Panglao. Sundays especially if the sea is nice, the people there could reach a few thousand sometimes. So a lot of trash is left behind. That's what we always do after weekends. We have to do a lot of cleanup. (L. Nunag, personal communication, October 12, 2023)

For Mr Nunag, the absence of communal or modular wastewater treatment facilities on Panglao Island greatly affects the smaller resorts in the Alona area. Larger accommodation establishments can afford their wastewater treatment facilities. However, the majority of resorts in Panglao Island are small, and micro-resorts only have septic tanks. He fears that Alona and other beach areas will continue to have issues with water quality under these conditions.

Mr Nunag also pointed out barriers to the problem of single-use plastic bottles in the destination. One is that no glass bottler or recycler for water is accessible to Bohol. On the consumer side, there is still a need for a more energetic campaign on responsible behavior for tourists and residents alike. For example, the practice of bringing one's water container still has not caught on with tourists.

While he is a strong advocate for sustainable procurement and supporting local products, Mr Nunag believes that the government needs a more explicit commitment to helping local farmers and MSMEs to become viable suppliers to the tourism industry. He cited the example of Morocco, which impressed him with its support program for their MSMEs to deliver culture- and tradition-based souvenirs that offer value for money despite the scale disadvantage.

In this vein, the local hotel and resort association has been lobbying with the provincial agriculture office to encourage more local farmers to cultivate fruits and vegetables that are still being imported from Bukidnon and other provinces. They have also been working with artists and handicraft makers to encourage more production of locally crafted souvenirs, sourced primarily from local materials.

Sustainable Tourism Initiatives in San Juan, La Union

San Juan is a municipality of about 40,000 residents in the province of La Union, about a 280-kilometer drive north of the Philippine capital city of Manila. Traditionally a fishing and farming community, San Juan has become known in the country as its "Surfing Capital of the North." Surfing was introduced to La Union as early as the 1970s when American military personnel on rest-and-recreation breaks would drive over from Subic Bay Naval Base and Clark Air Base.

However, it was not until the middle of the 2010s that La Union began to take off as a surf tourism destination. From an average of 30–35 thousand per year at the turn of the millennium, overnight visitor arrivals began to consistently reach 200 to 250 thousand arrivals by 2016. Almost 90% of these arrivals come for leisure and recreation, with San Juan's Urbiztondo surfing area as the most preferred attraction site in the province according to survey results shared by the La Union Provincial Tourism Office on its official Facebook page (Love La Union, 2022, July 15).

While tourism growth has been welcomed, local stakeholders have become wary of the problems that could arise from overtourism. Private business groups and civil society groups began organizing various initiatives for environmental preservation and conservation and advocating for local water conservation and solid waste management policies. They successfully lobbied for a series of provincial ordinances to regulate the use of single-use plastics and plastic drinking straws that were eventually merged and unified as Provincial Ordinance No. 342-2023, also known as "The Single-Use Plastics (SUPs) and Other Reusable and Disposable Materials Code Of The Province Of La Union." The private and community lobby was also instrumental in the enactment of Provincial Ordinance No. 088-2016 which provided for the creation of a water conservation committee for the province, as well as counterpart committees in all LGUs of La Union.

In line with all these, the San Juan Resort Restaurant and Hotel Association (SJRRHASS), which was originally established as a private sector promotional vehicle for the destination, shifted its mission in 2016 to include sustainable tourism practices in its community development goals (Institute for Solidarity Asia, 2022).

Under the umbrella of the La Union Soul Movement campaign, SJRRHASS has partnered with the local association of surfing instructors as well as various nonprofit groups on activities related to sea turtle conservation, plastic waste recycling to produce eco-bricks, and *Sustainable Palengke* – a campaign to encourage the use of reusable bags in place of single-use plastics in the public market. In 2018, the San Juan tourism stakeholders redesigned the three biggest festivals of the year to incorporate more green features and messaging. These included breaking up the festival events into smaller pocket events distributed across different locations around town – instead of the traditional program involving massive crowds in a single venue along the beach.

In an online interview posted on YouTube, La Union Soul convenor and SJHRRASS President Tina Antonio (2023) explained that the move toward dispersed, mini-events was meant to reduce the noise and bright evening lights which could disrupt the nesting patterns of sea turtles, locally known as *pawikan*. In 2021, the La Union Soul Movement joined the 3-year Clean Coasts Project, a multistakeholder undertaking spearheaded by the Coca-Cola Foundation and the

Jaime V. Ongpin Foundation.[7,8] The project aims to protect and preserve the coastal marine environment of La Union, by focusing on plastic recovery and recycling, innovative waste management systems, and community education (Antonio, 2023).

In a group interview conducted with the industry leaders directly involved in these projects, the stakeholders expressed positive feelings about how the different projects have gained them some tangible momentum in their vision of a sustainable San Juan. However, they acknowledge that the situation is still many years away from reaching a complete sea change in attitudes and behavior necessary for San Juan's long-term flourishing.

A sustainability consultant of the Provincial Government believes that the overconcentration of visitors and tourism investments in the San Juan area will continue to pose a threat to their ability to ensure good water quality, solid waste management, and conservation of the wetlands and watersheds in neighboring towns of the province. Redistributing tourism demand sustainably to these other destinations within La Union will be crucial in addressing these.

Ian Saguan, a surfing instructor and president of the La Union Surf Club, emphasized the importance of building positive environmental attitudes and habits. He lamented that casual visitors – those who did not come primarily to surf – did not share the surfer ethos that characterized "the surfer community" in San Juan. He feels that "regulars," i.e., surfers who keep returning to San Juan, have a stronger connection with the land and sea than those who only come for a taste of its surfer vibe but "don't have any concern for the environment."

While the general sentiment from stakeholders in San Juan is that small but positive steps are being taken, they also see the bigger picture concerning the province, especially with the neighboring destinations along the coast. They see the small victories in San Juan as catalysts for more positive environmental action beyond Surf Town. For example, Project CURMA or Coastal Underwater Resource Management Actions, the sea turtle conservation project in San Juan, has provided the movement with an entry point for engaging both residents and visitors in conservation action.[9] CURMA and its partners have been able to harness the initial curiosity and concern about the *pawikan* as an endangered sea creature, to hold deeper conversations about its importance to fishing (e.g., managing seagrasses necessary for fish propagation) and tourism (e.g., controlling the jellyfish population). From that vantage, the advocacy can be expanded to the conservation of the coastline and the ocean, i.e., the way to protect the sea turtles is to protect our shared environment. In the process, CURMA has created alternative sources of income for residents who used to poach the turtles for meat and eggs.

The San Juan environmental champions have also been very active in supporting efforts to preserve almost 80 hectares of mangroves in the town of Bauang, about an hour south of San Juan. Working with local communities in

---

[7]For information about the Coca-Cola Foundation: https://www.coca-cola.com/ph/en/social/coca-cola-foundation

[8]For information about the Jaime V. Ongpin Foundation: https://jvofi.org/

[9]For more information about CURMA: https://sifcare.org/curma/

three barangays, they have helped to establish the Bauang Bakawan Eco-tourism Park.

According to Ms Antonio, the activism of San Juan's stakeholders has been a testament to the power of collective action and the willingness to engage with public and private partners.

> We were able to stop a coal-fired power plant from being built in Luna (the neighboring town). It took us eight years. So that's progress. Now, we do have a lot of partnerships with government. We got more invites to attend planning activities with our government, and tourism partners. Just being able to get a voice is, I think, progress. (T. Antonio, 2023, personal communication, October 19, 2023)

The TTVC Project in the Philippines

Launched in 2017, the TTVC project sought to train hotels, restaurants, and MICE venues across the Philippines to adopt and replicate best practices for reducing negative environmental impacts, especially GHG emissions, along three areas: energy efficiency, water conservation, and waste management (solid and liquid). In 2019, the project assisted the DOT and the private sector Tourism Congress of the Philippines in developing the national "Roadmap for low-carbon and resource-efficient tourism" (UNEP, 2019).

One key output resulting from the TTVC project was a tool for assessing the resource efficiency and GHG emissions (RE tool) of tourism establishments over a 10-year frame. The project used the RE tool to assess almost 40 tourism businesses, including 13 hotels that were further assisted in analyzing the environmental impacts of their operations using individualized sustainability reports. TTVC also produced short case studies, as informational material, about how these hotels integrated sustainability, gender, and environmental performance monitoring into their operations.

June Alvarez was the executive director of the Philippine Center for Environmental Protection and Sustainable Development, Inc. (PCEPSDI), the nonprofit organization that implemented the TTVC project in the Philippines. In his assessment, the TTVC, as an environmental advocacy campaign, was somewhat successful in raising the appreciation of the industry for environmentally responsible operations, particularly within the hotel and resort sector. Mr Alvarez suggested that the hotels and resorts were generally responsive to their calls for participation because they also recognized the growing interest in sustainable consumption and production (SCP) from their markets. For example, hotels and resorts have been receiving positive feedback for things like energy efficient lighting and air-conditioning systems, pump dispensers for shampoo, soap, and lotion in the bathrooms, or replacing plastic water bottles with hallway water dispensers for guests to fill the reusable pitchers provided in their rooms. Restaurants and souvenir shops have also recognized the marketing value of farm-to-table, locally sourced and manufactured souvenirs, and similar products that visitors perceive not only as authentic but also carbon-friendly because of the reduced shipping and transport costs. Conversely, the hotels are also observing more guests with reusable shopping bags, drinking bottles, and even spoons, forks, and chopsticks as their own expressions of sustainable travel consumption.

However, he acknowledged that most of their partner establishments were large business establishments that could afford the cost of sustainability training for their staff. Even when the managers were convinced about the importance of committing to greener operations, most smaller properties could not as easily afford to pull staff from regular duties to participate in the TTVC training workshops.

One specific activity that the Project implemented in 2019 was to incorporate the UNEP-initiated "Campaign to phase out single-use plastics (SUP) in the tourism sector" into its program of activities, in line with the "Save our Spots" responsible tourism campaign that the DOT had launched around the same time. However, according to Mr Alvarez, eliminating plastic is very difficult without viable alternatives. He noted that the cheap, nonsustainable products and packaging remain conveniently available in the Philippines. In addition, manufacturers of local reusable or recyclable products do not have the economies of scale necessary to become viable producers of sustainable alternative products.

> The challenge I see is that if they use local materials, the economies of scale of production are missing. Our productivity is generally low in the Philippines. With globalization nowadays, they (consumers) can easily access materials and resources. At the end of the day, after they use it, it (plastic packaging and other waste) will end up in the environment. (J. Alvarez, personal communication, October 10, 2023)

This echoes observations from a UN-Habitat report (Oposa et al., 2022) that also looked at barriers to plastic waste reduction behavior change. It found that both consumers and retailers were challenged by the extra effort that reusables required, in contrast to disposable products. Retailers who tried to adopt a refill-based delivery option for products such as shampoo, cooking oil, soy sauce, etc., encountered problems in replenishing their stock. In addition, commercial brands preferred by their customers were not always available or suitable for the refill model. Ultimately, the financial sustainability of such business models depended on the ability to sustain responsible consumer behavior on a wider scale.

Observations

The three cases presented in this chapter all recognize the tourism industry as a chain of inter-related enterprises whose combined output forms a collective experience of the destination. In two cases, the initiatives were specifically branded as TVC interventions. In the case of San Juan, the stakeholders did not explicitly label their program with the TVC appellation. However, their approach remained consistent with key elements of TVC in terms of recognizing the supplier–buyer relationships among different types of tourism businesses and building partnerships from these networked interdependencies to collectively respond to industry and environmental concerns.

As industry-centric or industry-led initiatives, all three cases designed around helping tourism businesses, especially hotels and resorts, become more sustainable by reducing the negative environmental impacts of their operations. In both Panglao and San Juan, this process was facilitated by the presence of organized private sector associations which could mobilize their members to become more responsible operators. Active tourism associations also helped to create a climate of mutual support and knowledge sharing that encouraged more members, and perhaps influenced nonmembers, to follow the lead of the early adopters.

As the frontline of the tourism industry, the private tourism establishments can also be very effective in the information and education campaigns for responsible consumer behavior through their direct contact with visitors. By partnering with local environmental nongovernmental organizations (NGOs,) the tourism establishments provide platforms for the NGOs to directly reach visitors to the destination.

The cases of Panglao, San Juan, and the TTVC Project also demonstrated the utility of the value chain framework in identifying sustainability "hot spots" that priority efforts can be directed toward. In Panglao, this partly entailed equating sustainability with green practices that emphasized resource efficiency toward reducing a property's overall carbon footprint. These include the increased use of energy-efficient fixtures and equipment, the reduction or elimination of nonrecyclable/nonreusable supplies, and the strengthening of local supply chains to reduce dependence on imports.

However, because the TVC framework is based on value creation and business productivity, the identified hot spots and corresponding interventions were limited to the core operations of participating firms. As a result, there is a tendency for program leaders and participants to limit their ecological view of the TVC to the environmental impacts of their operations, rather than a comprehensive ecosystems-oriented appreciation of the situation.

While commendable, the value chain-based framework for sustainable tourism applied in the Philippine cases does not seem comprehensive enough for the more complex challenges of the destination. Part of the reason might be the rather limited theory of change that the case studies suggest.

In all three projects presented here, the advocacies for sustainable tourism have been anchored on the adoption of essentially new technologies by tourism establishments in given destinations. Thus, many aspects of the change processes that destinations such as Panglao and San Juan have undergone are consistent with models such as Venkatesh and Davis' (1996) Technological Acceptance Model (TAM). The TAM suggests that when presented with a new technology, the users' decision to use or adopt it is influenced by two things: its perceived usefulness and its perceived ease of use. External variables such as social influence, image, and other factors may affect the users' perceptions of the new technology, as well as their subsequent attitudes and intentions about adopting it.

In all three case studies, sustainable tourism has largely been communicated as new technology, i.e., new and better ways of reducing waste, improving energy efficiency, and minimizing ecological impacts from hotel and resort operations. These technologies were introduced to local establishments through technical assistance from external agents such as GTZ's Greening the Tourism Value Chain project in Panglao and UNEP's Transforming Tourism Value Chains project. In contrast, the catalyst for sustainable tourism practices in San Juan emerged almost organically from internal advocacy groups and personalities.

As shown in Fig. 3.2, the first recipients of the technical became the early adopters or pioneers of sustainable tourism in their respective locations. Interestingly in Panglao, San Juan, and the TTVC project, the pioneer establishments generally belonged to the accommodations sector. As Mr Nunag in Panglao suggested, the extent to which the pioneers or early adopters were able to sustainably demonstrate the efficacy and cost-effectiveness of the sustainable technologies played a big part in convincing other hotels and resorts to follow their

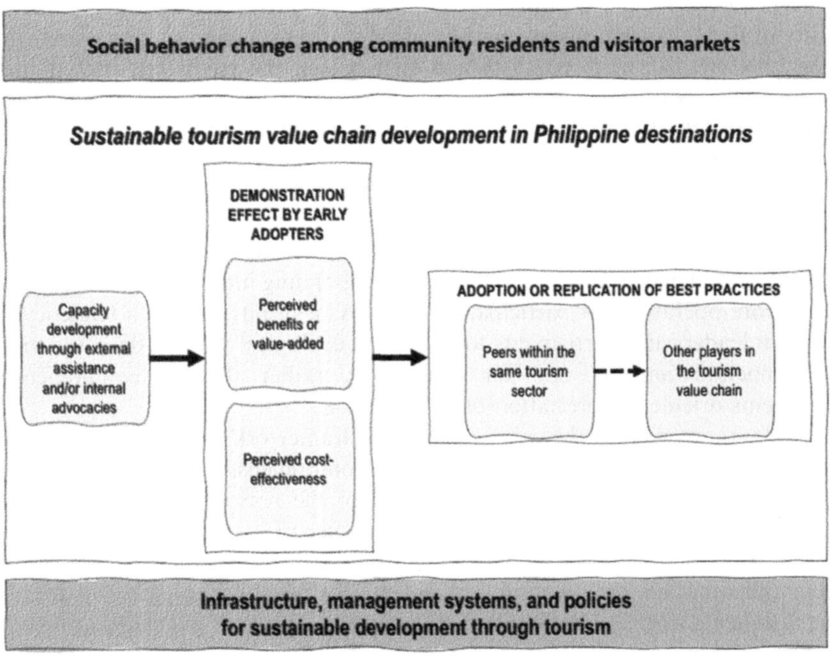

Fig. 3.2. Tourism Value Chain Interventions to Catalyze Sustainable Tourism Development. *Note:* Author constructed.

examples. Eventually, other types of businesses in the TVC may be positively influenced by hotels and resorts demanding more sustainable goods and services from local producers and suppliers.

The San Juan experience followed the same general approach as Panglao and TTVC in highlighting best practices for players in the local value chain to emulate. However, San Juan has been able to offer a much more diverse group of early adopters – perhaps because the push for sustainability came internally, from a wider base of stakeholders. Thus, San Juan's LU Soul Movement includes not only hotels and resorts but also restaurants, souvenir-makers, and other types of businesses.

Nevertheless, all informants interviewed agreed that their respective programs, given their common focus on improving the TVC, addressed only a limited side of the sustainability puzzle. In all three case studies, the informants acknowledged two areas of concern that their value chain models did not address sufficiently: ensuring the presence of the enabling systems and infrastructure needed and effecting demand-side behavior change from their clients as well as from the residents of their communities.

One reason for this shortcoming might be that Philippine value chain-based programs have largely discounted an active role for local government relative to the chain of business activities in the destination. The experiences of Panglao, San Juan, and other destinations suggest that local governments should not merely be regulators but also actors in fostering sustainable development and tourism. The consequences of overdeveloping tourism sites at the expense of natural areas (such as Panglao's beach areas and the mangroves and wetlands around San Juan) require large infrastructure investments and complex management systems that often exceed the capabilities of individual players from the private sector or local community.

One specific gap relates to the infrastructure and public services needed to consolidate and scale up the gains from the TVC-based projects. The success stories of sustainable solid waste management from Panglao's Greening the Tourism Value Chain, San Juan's LU Soul movement and the nationwide TTVC Project were all limited to micro or firm level practices. However, for these individual efforts to be scaled up to the community or destination-level, local governments need to ensure that the destination-wide logistics for effective waste collection, processing, and disposal are in place so that these individual efforts can be extended more widely to their respective communities.

Local governments also play an important role in empowering smaller enterprises to adopt more sustainable business operations. Micro and small enterprises often do not have the resources that large firms have to engage with suppliers of green technologies, products, and services. Sustainable production and procurement can be especially challenging for MSMEs because the cost of shifting to more sustainable operations can be prohibitive for them.

Local governments can strategically direct their incentive policies and programs to encourage more investments in renewable energy generation, recycling, and other shared facilities that can make these services more accessible and affordable to the smaller players in the local economy.

They will also need to engage the industry at many different points to increase the proportion of locally produced inputs into the local tourism product. This can include an explicit commitment to boost the productivity of local suppliers, especially those that can deliver greener, less wasteful inputs to tourism services. As observed by Philippine Center For Environmental Protection and Sustainable Development, Inc. (PCEPSDI) and UN-Habitat (Oposa et al., 2022), no matter how strongly the tourism industry argues for local sourcing, community-based suppliers are generally not productive enough to offer competitive options over a sustained period.

Most tourism establishments will find it difficult to ignore overall convenience and practicality of mass-produced imports. Artisanal crafts and other products manufactured using local materials will not be able to compete on price against the mass-produced, disposable alternatives assistance from government. Local governments could provide common service facilities to help small players and their suppliers gain access to the technology needed to be more productive and competitive against the mass-produced and often wasteful imports. Logistical and technical support to improve product design and manufacturing quality standards of micro and small manufacturers can help them deliver more competitive value offers – particularly when combined with compelling narratives of place and product.

Agriculture can be the low-hanging fruit that Philippine destinations can start from in building up their TVCs to become more sustainable and inclusive. Many of the country's established tourism destinations, such as Bohol and La Union, are based on coastal, marine, and similarly nature-based environments that are also likely to have rural economies. Deliberately extending the supply chains back toward local farmers and fishers can not only shrink these destinations' carbon footprints. It can also reduce their economic leakages and enable more community members to share in the retained economic benefits of tourism. Finally, strengthened supply links between tourism businesses and agricultural producers may create opportunities for diversifying the destination's tourism product through food, health-and-wellness, and other farm-related tourism activities. These may then allow for tourism to be dispersed beyond the overconcentrated beach resort districts and city centers toward the peripheral communities.

Conclusions: The Value Chain as a Framework for Sustainable Tourism

Given its origins as a framework for assessing firm-level issues of productivity and competitiveness, value chain analysis has naturally found more usage related to the economic activities and impacts associated with the value creation process of a tourism industry. TVC analysis has also been employed in developing strategies to improve destination competitiveness or to identify opportunities for the constraints faced by low-income and other disadvantaged groups in gaining access to livelihood and job opportunities from tourism.

However, as suggested in Fig. 3.3, value creation process in tourism more often flows vertically – along supply chains for hotels, restaurants, transportation, etc. – and not horizontally from one subsector to another. Thus, the destination-based as

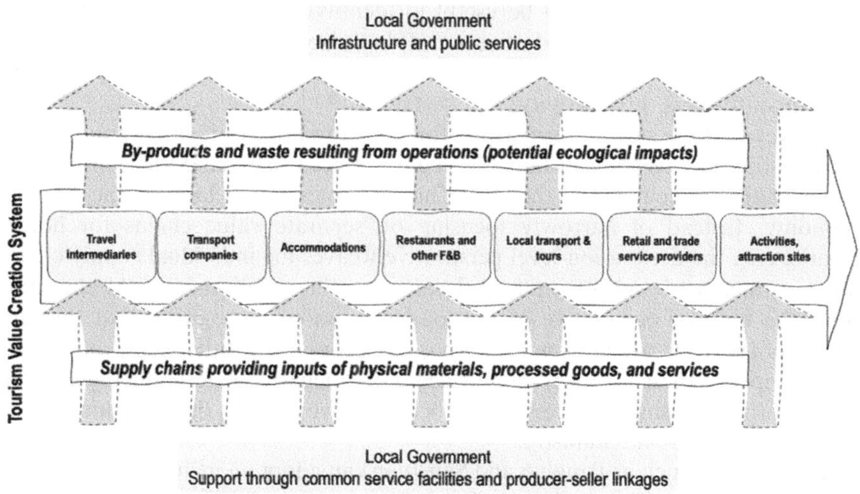

Fig. 3.3. The Tourism Economy as a Web of Supply Chain Inputs and Value-Creating Processes. *Note:* Author constructed based on Partale (2020).

well as national value chain projects highlighted in this chapter generally designed their interventions around specific types of establishments, i.e., hotels, resorts, restaurants, and convention facilities, rather than the larger destination or local economy.

Concentrating on specific types of tourism establishments may be more efficient for purposes of project implementation, especially in the donor-driven cases of Panglao and the TTVC project. However, such approaches risk minimizing or ignoring the much more complex web of relationships on which tourism destinations are built. The services that make up the tourism chain do still require physical inputs from other suppliers. These service businesses do require water and energy in their operations. Finally, these tourism businesses also contribute physical waste to whatever location or destination they are in.

The experiences of destinations such as Panglao, San Juan, and Palawan (in the case of the TTVC project) provide important lessons for tourism managers and policymakers. The first is that sustainable tourism development starts from the bottom-up. Given how geographically dispersed and physically diverse are tourism destinations across the country, national campaigns for sustainable tourism such as the Transforming Tourism Value Chains Project will be more effective in the aspects of areas of enabling policy and in general awareness-building among consumers, host communities, and tourism operators.

It is at the destination level where the practicalities of environmental, social, and economic sustainability must be dealt with on a day-to-day basis. The TVC approach has been shown to be useful in identifying the initial "hot spots" or areas of work in a particular destination. It has also provided a starting point for rallying and organizing the local private tourism sector around unifying concerns of competitiveness, productivity, and sustainability.

However, if the stakeholders aspire for transformative outcomes that establish their destinations as truly sustainable, there is a need to expand the firm-oriented value chain framework to the more holistic, meso-level lens of the tourism economy. Instead of narrowly focusing on separate value chains for hotels, resorts, etc., the destination-level perspective weaves the individual supply chains into an integrated web of input and value-creating relationships. At the same time, the broader perspective of the tourism economy recognizes that the outcomes of the TVC are not limited to positive economic results, i.e., productivity and competitiveness. It also acknowledges the negative externalities generated from tourism's business processes and their negative impacts on communities and ecosystems outside the industry.

Destinations such as Panglao and San Juan can adopt an inclusive approach to sustainable tourism that builds out from the core of the TVC to include supplier industries and communities into a whole-of destination model of tourism development. This can only be accomplished from a partnerships-based paradigm of local development that facilitates synergies among the disparate efforts of public, private, and civil society groups. Often, the local government will be the one stakeholder group that has the motivation and capability to maintain a holistic perspective of the situation in the destination.

Thus, the local authorities are the stakeholders best positioned to coordinate and help scale up the initial efforts of private and civil society groups toward a harmonized program of sustainable transformation.

Among the key things that local leaders can do are:

- Institutionalize the practice of convergence and public–private partnership-building within the tourism industry and across different economic and social sectors.
- Investing in continuous information, education, and capacity-building to empower different stakeholder groups to participate effectively in the destination's transformation process and continuing the theme of inclusive engagement.
- Prioritize financial, technical support, and capacity development for MSMEs to take part in the partnerships for sustainable development and tourism, while continuing to support industry associations and sustainability champions to initiate or lead programs that they have the resources for.

Finally, sustainable destination development cannot depend solely on greening and transforming TVCs. Destinations must also acknowledge that their long-term success will also depend on being able to effect behavior change for sustainable consumption from stakeholders other than industry – visitors and, most importantly, residents.

References

Antonio, T. (2023, April 1). Plastic recovery and recycling for clean coasts program in La Union, Philippines. LinkedIn. https://www.linkedin.com/pulse/plastic-recovery-recycling-clean-coasts-program-la-union-tina-antonio/

Antonio, V. (2012). Local and Regional Economic Development (LRED) Toolkit Version 1.0. A participatory approach for building sustainable local economies. In R. Carpio (Ed.), *Deutsche Gesellschaft für Internationale Zusammenarbeit (GIZ) GmbH*. https://www.local2030.org/library/147/Local-and-regional-economic-development-LRED-toolkit-A-Participatory-Approach-for-Building-Sustainable-Local-Economies.pdf

Arbon, E. (2014). *Reporting ProGED accomplishments. Province of Bohol (01 January to 22 August 2014)*. [PowerPoint slides]. https://www.slideserve.abrielriel-valdez/reporting-proged-achievements-province-of-bohol-1-january-to-22-august-2014-ms-ma-elena-arbon

Arnaldo, S. (2023, November 28). South Koreans top foreign visitor arrivals in Bohol. *Business Mirror*. https://businessmirror.com.ph/2023/11/28/south-koreans-top-foreign-visitor-arrivals-in-bohol/

Aw, E., & Phuttitarn, L. (2020). *Market report. Bohol resort market, Philippines*. https://cdn.horwathhtl.com/wp-content/uploads/sites/2/2020/01/MR_Bohol-Resort-Market.pdf

Businessworld. (2019, January 21). Other tourism destinations learn from Boracay's closure. *BusinessWorld*. https://www.bworldonline.com/economy/2019/01/21/210169/other-tourism-destinations-learn-from-boracays-closure/

Cave, J., & Dredge, D. (2018). Reworking tourism: Diverse economies in a changing world. *Tourism Planning and Development*, *15*(5), 473–477. https://doi.org/10.1080/21568316.2018.1510659

Du Plessis, C. (2017). Going beyond sustainability: Changing views, changing ways. In P. S. Brandong, P. Lombardi, & G. Q. Shen (Eds.), *Future challenges in sustainable development in the built environment* (pp. 40–54). John Wiley & Sons, Inc.

Finkel, T., Duhaylungsod, N., & Bauer, S. (2010). *ProGED climate change and private sector development: Integrating green growth strategies into the MSME Development Plan 2010–2016. Deutsche Gesellschaft für Internationale Zusammenarbeit (GIZ) GmbH*. https://www.greenpolicyplatform.org/research/climate-change-and-private-sector-development-integrating-green-growth-strategies-msme

Goodwin, H. (2017). *The challenge of overtourism* (Responsible Tourism Partnership Working Paper 4, Issue October). https://haroldgoodwin.info/wpcontent/uploads/2020/08/rtpwp4overtourism012017.pdf

Gutierrez, N. (2023, July 10). This 5-Star Bohol resort is raising giant clams for sustainable eco-tourism. *Metro Style*. https://metro.style/travel/highlights/bellevue-hotel-bohol-sustainability/35900

Hummel, J., Gujadhur, T., & Ritsma, N. (2013). Evolution of tourism approaches for poverty reduction impact in SNV Asia: Cases from Lao PDR, Bhutan and Vietnam. *Asia Pacific Journal of Tourism Research*, *18*(4), 369–384. https://doi.org/10.1080/10941665.2012.658417

Institute for Solidarity in Asia. (2022). *Surf, sand and sustainability: How La Union catches the green wave through sustainable tourism*. https://isacenter.org/surf-sand-and-sustainability-how-san-juan-la-union-catches-the-green-wave-through-sustainable-tourism/

Ioannides, D., & Gyimóthy, S. (2020). The COVID-19 crisis as an opportunity for escaping the unsustainable global tourism path. *Tourism Geographies*, *22*(3), 624–632. https://doi.org/10.1080/14616688.2020.1763445

Jovicic, D. Z. (2016). Key issues in the conceptualization of tourism destinations. *Tourism Geographies*, *18*(4), 445–457. https://doi.org/10.1080/14616688.2016.1183144

Lara-Morales, O., & Clarke, A. (2022). Sustainable tourism value chain analysis as a tool to evaluate tourism's contribution to the sustainable development goals and local Indigenous communities. *Journal of Ecotourism*, 1–19. https://doi.org/10.1080/14724049.2022.2147189

Love La Union. (2022, July 15). Urbiztondo surfing area of San Juan remains to be the most enjoyed tourist attraction in the province. [Image attached] [Status update] Facebook. https://www.facebook.com/photo/?fbid=1844930402527326&set=pb.100069090378624.-2207520000

Oposa, A. R., Perello, H. S., Hibino, K., & Lavtizar, V. (2022). *Community behavior change strategies for plastic waste reduction*. UN-Habitat Regional Office for Asia and the Pacific.

Partale, K. (2020). *The tourism value chain. Analysis and practical approaches for development cooperation projects*. https://www.switch-asia.eu/site/assets/files/2460/giz_tourism_value_chains_en.pdf

Pearce, D. G. (2014). Toward an integrative conceptual framework of destinations. *Journal of Travel Research*, *53*(2), 141–153. https://doi.org/10.1177/0047287513491334

Pilarca, R. (2008). Accessing markets through the value chain approach. *The Private Sector Promotion Program (SMEDSEP)*. https://www.yumpu.com/en/document/view/46242911/accessing-markets-throughthe-value-chain-approach-smedsepph

Ramirez, R. (2018, November 24). Tourism establishments in Panglao next to be shut down. *The Philippine Star*. https://www.philstar.com/headlines/2018/11/24/1871263/tourism-establishments-panglao-next-be-shut-down

Raworth, K. (2017a). *Doughnuts economics: Seven ways to think like a 21st-century economist*. Chelsea Green Publishing.

Raworth, K. (2017b). Why it's time for doughnut economics. *IPPR Progressive Review*, *24*(3), 217–222. https://doi.org/10.1111/newe.12058

Romero, I., & Tejada, P. (2011). A multi-level approach to the study of production chains in the tourism sector. *Tourism Management*, *32*(2), 297–306. https://doi.org/10.1016/j.tourman.2010.02.006

Sharpley, R. (2020). Tourism, sustainable development and the theoretical divide: 20 years on. *Journal of Sustainable Tourism*, *28*(11), 1932–1946. https://doi.org/10.1080/09669582.2020.1779732

Sheldon, P. J. (2022). The coming-of-age of tourism: Embracing new economic models. *Journal of Tourism Futures*, *8*(2), 200–207. https://doi.org/10.1108/jtf-03-2021-0057

SNV. (2012). *Tourism and poverty reduction in Lao PDR: A reflection of SNV's work 2003–2011*. SNV Lao PDR Office.

Song, H., Liu, J., & Chen, G. (2013). Tourism value chain governance: Review and prospects. *Journal of Travel Research*, *52*(1), 15–28. https://doi.org/10.1177/0047287512457264

Spenceley, A., & Meyer, D. (2012). Tourism and poverty reduction: theory and practice in less economically developed countries. *Journal of Sustainable Tourism*, *20*(3), 297–317. https://doi.org/10.1080/09669582.2012.668909

UNEP. (2019). *Roadmap for low-carbon and resource-efficient tourism in the Philippines*. United Nations Environment Programme. https://www.oneplanetnetwork.org/sites/default/files/roadmap_for_low-carbon_and_resource-efficient_tourism_in_the_philippines_0.pdf

Vahlhaus, M., & Macaranas, F. M. (2005). Improving the business environment in the Philippines: A case study GTZ's support for a better business environment for small and medium-sized enterprised in the Philippines. *Presented at the International Conference on Reforming the Business Environment: From Assessing Problems to Measuring Results*, Cairo, 29 November to 1 December 2005. https://www.enterprise-development.org/wp-content/uploads/Valhausdoc.pdf

Venkatesh, V., & Davis, F. (1996). A model of the antecedents of perceived ease of use: Development and test. *Decision Sciences*, *27*(3), 451–481.

World Tourism Organization. (2018). *Tourism for development – Volume I: Key areas for action*. UNWTO. https://doi.org/10.18111/9789284419722

Chapter 4

Situating the Informal Sector in the Philippine Tourism Value Chain: Evidence, Opportunities, and Contentions in the Case of Binondo, Manila

Jovito Jose P. Katigbak[a] *and April Joy E. Dopeño*[b]

[a]De La Salle University, Philippines
[b]Far Eastern University, Philippines

Abstract

Informality in the tourism sector of the Philippines is a pervasive phenomenon, yet only few studies have explored the former's contribution to the latter, especially at the local level. Hence, this chapter utilizes a tourism value chain model developed by the United Nations World Tourism Organization and Directorate-General for Development and Cooperation (DEVCO) to determine the presence and examine the role of the informal sector in various spheres of Binondo's tourism industry. Findings reveal that informality is thriving in the areas of transportation, food and beverage, handicraft, leisure, excursion, tours, and support services. Conversely, they are inexistent or less visible in the accommodation and tourism assets in destination segments. Moreover, the fragmented tourism value chain of Binondo is heavily shaped by three primary challenges, namely, difference in perception between decision-makers and economic actors, resistance to change by the informal sector, and lack of relevant data to support evidence-based policymaking. To address these issues, the local government of Manila City may hence consider the following policy options: (i) mainstream a "whole-of-locality" approach, (ii) conduct an industry mapping of Binondo's tourism value chain, (iii) facilitate linkages between formal and informal businesses, (iv) develop shared infrastructure and extend capacity-building to the informal sector, and (v) employ a regulatory sandbox approach.

Revisiting Sustainable Tourism in the Philippines, 67–84

Published under exclusive licence by Emerald Publishing Limited
doi:10.1108/978-1-83753-678-820241004

Keywords: Informal sector; tourism value chain; informal tourism; Philippine tourism; Binondo tourism

Introduction

The informal sector employed 56% of the Philippine labor force or 22.7 million Filipinos, in 2017, and contributed a total of over PHP5 trillion to the country's gross domestic product (GDP) in 2016, according to Gonzales (2018) and Royeca (2023), respectively. Accordingly, the tourism industry has mainly benefitted from the goods and services offered by various sari-sari stores, street vendors, and pedicab drivers, among others. Nevertheless, workers in the informal sector are still subjected to low incomes, precarious employment, and inadequate or lack of social protection despite their role as a vital growth engine of the economy (International Labor Organization, 2023a). Hence, there is a need to support and enable the informal sector to effectively foster a more conducive path toward sustainable tourism in the better normal. This chapter thus seeks to locate the informal sector in the tourism value chain in Binondo, Manila, to aptly determine opportunities and outline policy options for key stakeholders in the Philippine tourism industry.[1]

Using the United Nations World Tourism Organization Tourism (UNWTO) Value Chain Model as reference, it attempts to identify the participation of the informal economy throughout the value chain process by reviewing secondary data. The current policy and legal landscape were similarly examined. The findings then served as inputs in developing policy considerations and solutions for the key stakeholders of the tourism industry in Binondo, Manila. These are characterized as opportunities since they are expected to be paramount in the country's pursuit of sustainable tourism post-COVID-19. The concluding section highlights issues, challenges, and sources of tensions involving the informal sector, formal sector, and the government. They are considered as roadblocks toward an integrated Philippine tourism value chain.

The insights and takeaways from this chapter may be beneficial on two prongs, namely, knowledge and policy. On the first component, this study contributes to the existing literature by providing a perspective on the role of informal economy in the Philippine tourism value chain, with a focus on Binondo, Manila. Previous studies have looked at the cited topic through the pro-poor tourism lens. On the second component, the key findings may serve as inputs for local government agencies in reviewing policies and formulating strategies to address issues faced by the informal sector and to maximize their participation in the value chain.

---

[1]Binondo is an administrative district of Manila comprised of 10 barangays. It is situated in the third legislative district of Manila along with Quiapo, San Nicolas, and Santa Cruz. In 2015, the 66-hectare Binondo estate hosted almost 13,000 individuals therefore having the distinction as one of the densest districts of the city (Boncan, 2015). It was spatially designed to contain the Christian Chinese (*Sangley cristiano*) during the Spanish occupation.

Informality and Tourism: The Philippine Context

The International Labor Organization (ILO) (2002) defines the informal economy as "all economic activities by workers and economic units that are – in law or practice – not covered or insufficiently covered by formal arrangements" (p. 25). Simply, they are operating outside the legal scope of the law. Enterprises and firms in the informal economy are also not covered in practice; thus, the law is not effectively enforced over them. Further, compliance with the law is discouraged due to outdated, cumbersome, and costly legal measures (ILO, 2002). These predicaments consequently result in absence of appropriate labor legislation, inadequate social protection and security, poverty, social exclusion, and heightened vulnerability. Moreover, work in the informal economy is often associated with cramped workplaces and nonfixed premises, low skills due to lower level of education, insufficient or irregular income, poor working conditions, extended working hours, and lack of access to markets, financing, information, technical know-how, and technologies (ILO, 2023a). Hence, Chen (2016) posits that mainstream thinking perceives the informal economy as "illegal and non-productive, as a symptom of bad governance and a drag on economic growth" (p. 27).

Informal employment can be characterized as "as any activity of persons to produce goods or provide services for pay or profit that is not effectively covered by formal arrangements such as commercial laws, procedures to report economic activities, income taxation, labor legislation and social security laws and regulations providing protection against economic and personal risks associated with carrying out the activities" (ILO, 2023b). Workers may be categorized into three major groups, namely: (i) employees, (ii) employers (with hired workers) and own-account workers (without hired workers), and (iii) contributing family members. The first type refers to individuals who do not receive social security contributions from their employer nor are entitled to paid annual or sick leave. The second group are business owners who run a nonincorporated, unregistered private enterprise with five employees or fewer and/or without a formal bookkeeping system and a fixed workplace. The last type encompasses unpaid family workers in own family-operated farm or business (ILO, 2018).

In the Asia and the Pacific, ILO (2023a) estimates that the informal economy employs 66% of all workers and is buoyed by 84% of all economic units as of 2019. The services sector comprises 41% of the total informal employment figure, followed by the industrial (30%) and the agricultural (29%) sectors. In addition, informal employees in the rural areas account for 85% of the labor force in the Asia and the Pacific as of 2019 (ILO, 2023a). Informal workers are usually males, 15–24 years old, 65 years old and above, and with low levels of education.

At the domestic level, there is no standard definition of informal employment, but informal workers can be classified into three primary classes: (i) self-employed without paid employee, (ii) wage and salary workers in precarious employment, and (iii) unpaid family workers in own family-operated farm or business (Gonzales, 2018). These workers are usually vulnerable, engaged in short-term, seasonal jobs or businesses, and work for different employers on a daily/weekly basis. From 2008 to 2017, the informal economy hired an average of 56% of the total Philippine labor force or 21.1 million Filipinos (Gonzales, 2018). Of the latter figure, the first category composed 51% (10.8 million) of the total, followed by the second type at 31% (6.5 million), and the last classification at 18% (3.8 million) during the same period.

Notably, 44.4% of informal workers, on average, are in the agriculture sector from 2008 to 2016. The services sector ranked second at 43.6%, while the industrial sector has the least number of laborers at 12% during the same period (Gonzales, 2018). Further, 60% of the workers are males, while the females account for 40%, on average, from 2008 to 2017. Chongvilaivan and Chooi (2021) estimate that the informal sector comprises 28% of the country's GDP, and this results in a narrow tax base since they exhibit tax avoidance behaviors coupled with weak tax enforcement and administration. Due to the prevalence of informal workers in the services sector, it is thus worth examining and describing the former's position within the country's vibrant tourism industry.

The Philippine Tourism Industry in Figures

The rapid expansion in the hotel, resort, casino, and gaming industries along with the continued growth in international cruise and nautical tourism in recent years has contributed to greater demand for tourism and travel-related services in the country (Caynila et al., 2022). It also ranked eight globally as a favored destination for medical tourism (Shead, 2017). Thus, travel services accounts for around 20% of the total services exports in 2019 and is regarded as one of the primary sources of foreign exchange inflows (Caynila et al., 2022). It amounted to USD9.8 billion in 2019, and tourism receipts comprised 95% of the cited figure.

Locally, the Philippines' tourism direct gross value added (TDGVA) increased from about PHP851 billion in 2012 to PHP1.3 trillion in 2022 according to the Department of Tourism (DOT) (2023a). However, this corresponds to a 6.2% share to the country's GDP in 2022, which is lower than the TDGVA's 7.7% share to GDP in 2012.[2] Caynila et al. (2022) underscored that better tourism infrastructure such as airports, land transport, and hotels during the same period catalyzed the industry's growth. Moreover, the gross value added of tourism industries (GVATI) was measured at PHP3.3 trillion, which equates to 16.9% of the country's GDP in 2019 (Caynila et al., 2022).[3]

The preliminary results of the Philippine Statistics Authority's (PSA) (2023) 2020 Survey of Tourism Establishments in the Philippines find that there were 40,815 tourism establishments in 2020 compared with 38,699 in 2014. Food and beverage serving activities had the largest share to total establishments at 60.2%, followed by education, health and personal service activities (18.6%), and accommodation for visitors (10%). The top revenue-generating industries in 2020 are food and beverage

[2]TDGVA is defined as the total gross value added by tourism industries of the economy that directly serve visitors in response to internal consumption (Caynila et al., 2022).
[3]GVATI refers to the sum of all gross value added from establishments that are considered tourism industries, regardless of whether they serve tourists or not (Caynila et al., 2022).

serving activities with PHP188.4 billion, air passenger transport with PHP58.8 billion, and accommodation for visitors with PHP40.4 billion (PSA, 2023). Regarding employment, the sector hosted 5.3 million workers in 2022, from 4.5 million in 2012 (DOT, 2023a). The former figure represents 11.4% of the country's total employment, while the latter amounts to 12.1%, which illustrates a minimal decline.

Notably, the year 2020 was a grim year, to say the least, for the tourism industry due to the advent of the COVID-19 pandemic which brought border shutdowns, mobility restrictions, and strict quarantine restrictions across the world (Caynila et al., 2022). International tourist arrivals plummeted by 74% during the same year which translates to an estimated loss of USD1.3 trillion receipts (UNWTO, 2021). Domestic tourism was also adversely impacted by the enhanced community quarantine and risk-averse tourists which caused a 61.2% contraction in TDGVA and a 72% decline in GVATI in 2020 (Caynila et al., 2022). Due to the closure of many tourism establishments, employment decreased by 18.1% during the same year, especially in the food and beverage, accommodation, and passenger transport sectors. Given the ubiquity of the informal sector in the tourism and travel-related services, it can be surmised that informal workers were heavily affected by the economic squeeze caused by the COVID-19 pandemic.

To support the tourism industry in its postpandemic recovery, the DOT modified its Tourism Response and Recovery Plan (TRRP) by outlining "new normal" measures on travel including sustained business operations, updated infrastructure policy, better marketing and market and product development, and the development of guidelines on the health and safety of its stakeholders (Caynila et al., 2022). The TRRP likewise identifies both fiscal and nonfiscal measures for tourism establishments as they navigate a post-COVID-19 economic landscape. Approved by President Ferdinand Marcos Jr (PBBM), the DOT presented the completed but yet to be finalized National Tourism Development Plan (NTDP) 2023–2028. It aims to "establish a Philippine Tourism Industry anchored on Filipino culture, heritage, and identity, which aims to be sustainable, resilient, and competitive in order to transform the Philippines into a tourism powerhouse in Asia (DOT, 2023b)." Moreover, the NTDP features seven strategic goals, namely: (i) enhanced tourism infrastructure and accessibility, (ii) cohesive and comprehensive digitalization and connectivity, (iii) improved overall tourist experience, (iv) equal tourism product development and promotion, (v) promotion of a diversified, multidimensional tourism portfolio, (vi) better domestic and international tourism, (vii) strengthened tourism governance via close collaborations with national and local stakeholders (Presidential Communications Office, 2023).

More importantly, Chapter 7 of the Philippine Development Plan (PDP) 2023–2028 outlines the government's strategies in reinvigorating the services sector, which is expected to result in three main outcomes, namely, market expansion, stronger creativity and innovation in services value proposition, and better inter-sectoral linkages (National Economic Development Authority (NEDA), n.d.). The tourism sector is identified as a substantial contributor to the first outcome hence the need to advance resilient, sustainable, and inclusive multidimensional tourism, enhance the quality of the Philippine tourism experience, and actively promote tourism and culture. The DOT is mandated to take the lead in achieving the cited objectives through the implementation of relevant frameworks and programs such as the NTDP 2023–2028, Barrier-Free Tourism, Anahaw Philippine Sustainable Tourism Certification, Philippine Tourism Human Capital Development Plan 2021–2025, and National Cruise Tourism Strategy, among others (NEDA, n.d.).

For its part, the Department of Labor and Employment (DOLE) oversaw the implementation of the National Employment Recovery Strategy (2021–2022) which identified an eight-point action plan toward the road to recovery of local businesses and workers (Gonzales, 2022). More particularly, points six and seven somewhat cover the informal sector through the extension of financial assistance and social protection to vulnerable groups. Specific projects are the DOLE Integrated Livelihood Program, Tulong Panghanapbuhay sa Ating Disadvantaged/Displaced Workers, Barangay Ko, Bahay Ko (TUPAD #BKBK), Zero Hunger, and the Balik Probinsya, Bagong Pag-Asa Program (Gonzales, 2022).[4,5] The Department of Social Welfare and Development likewise delivered livelihood assistance grants to sustainable livelihood program beneficiaries, who have at least one informal worker in the family who was displaced by the lockdown measures (Dreyer & Nygaard, 2020).

Situating the Informal Sector in the Tourism Value Chain: A Focus on Binondo, Manila

The significance of the informal sector in the tourism industry has already been explored by various authors. For instance, Lv (2020) found that the size of the informal economy is reduced during the nascent stages of tourism development then expands vis-à-vis tourism growth according to the data across 96 countries over the period 2000–2007. In Africa, Slocum et al. (2011) examined the informal tourism sector in Tanzania and cited the group's major impediments such as cumbersome bureaucratic procedures, weak marketing knowledge, duplication of items, limited credit sources, and overdependence on foreign aid. Nevertheless, the informal tourism sector is commended for creating, negotiating, and executing codes of conduct on selling that all vendors should abide by. The role of local governments in developing the mentioned sector is paramount as evidenced by the case of Johannesburg, South Africa, which has become a shopping hub of sub-Saharan Africa (Rogerson, 2018).

In Asia, Truong (2018) delved into the experiences of street vendors engaged in tourism in Hanoi, Vietnam, and found that majority of them came from the rural

[4]TUPAD #BKBK is a community-based package of assistance that provides temporary wage employment for the underemployed, self-employed workers, and displaced marginalized workers who lost their livelihood or whose earnings were affected due to the Enhanced Community Quarantine against COVID-19. For more details, see: https://www.dole9portal.com/bkbk/

[5]This program aims to decongest Metro Manila's urban areas by encouraging households, especially informal settlers, to return to their home provinces while receiving government support and incentives on transportation, livelihood, housing, subsistence, and education, among others. For further details, see: https://bp2.nha.gov.ph/

areas due to low farm productivity and inadequate nonagricultural jobs. Aside from the meager income, informal workers are challenged by the government's ban on street vending issued in 2006 since they are viewed as negative representation of tourism and a peculiar characteristic of cultural heritage. Vendors who were proven to have approached or invited tourists to their stalls are obliged to pay a fine, surrender their commodities, and issue a letter committing to cease their street vending (Truong, 2018).

In Chiang Mai, Thailand, Cakmak et al. (2018) noted that the informal tourism entrepreneurs offer an array of products including handicrafts and semi-finished items that are supplied to other manufacturers (e.g., wooden cutlery) and hotels and catering firms (e.g., meals, vegetables). They also provide services related to transportation, entertainment, tours, and accommodation. Interestingly, informal tourism entrepreneurs in Chiang Mai emphasized the significance of greater collaboration with formal actors and public sector organizations, especially on areas related to designing and packaging, marketing, capacity-building, financial support, social capital, and intellectual property rights of cultural capitals. Williams and Shaw (2011) have already underlined the significance of partnerships between the formal and informal sectors as it may foster a more diverse and resilient supply chain for the former, while Zhang et al. (2009) stressed the cultivation of inclusive tourism value chains. Formal enterprises may also relish trust and appreciation from local communities and customers due to their expanded social responsibility through tie-ups with the informal sector (Park et al., 2014).

In the Philippines, several studies have touched on the spatial reconfiguration caused by gentrification (Back, 2019), decision-making dynamics (Malogdos & Yujuico, 2015; Milgram, 2014), and gender relations (Carlos et al., 2023). Rivera and Gutierrez (2018) developed a framework for ecotourism value chain and argued for a "more concerted multistakeholder approach to sustainability with enhanced government participation in creating a business environment practicing the quadruple bottom-line approach" (p. 12).[6] They likewise highlighted the role of government and big businesses in capacitating small firms through access to financing, nonmonetary incentives, resources, facilities, technologies, and technical know-how. In another study, One Planet Network (n.d.) revealed that the primary production of food, electricity, and fuel use in hotel and meetings, incentives, conferences, and events (MICE) establishments, and liquid and solid waste generation by hotel and MICE establishments are environmental hotspots in the country's tourism value chain.[7] Notwithstanding the works of these authors, it can be observed that there is scant literature on the application of a tourism value chain approach at the local level. Hence, this chapter endeavors to locate the informal tourism industry in the vicinity of Binondo, Manila, which is a renowned tourist destination for foreigners and locals alike.

---

[6]This pertains to sustainable tourism anchored on: financial growth, environmental stewardship, community engagement, and organizational development. See discussion by Rivera and Gutierrez (2018).

[7]An environmental hotspot analysis attempts to pinpoint areas in a value chain that have significant adverse environmental impacts.

Profile of Binondo, Manila

Today, like before, Binondo is widely regarded as the centerpoint of Chinese commerce in Manila as evidenced by numerous shops and stalls offering a whole spectrum of goods and services to every consumer. As aptly described by Boncan (2015),

> . . .it is commerce that requires the daily convergence of thousands of people in every street, nook, and cranny of Binondo who come to buy, to sell, to carry out professional services, to provide food, to transport goods, services, and people. (p. 124)

She further stressed that the sprawling and still growing populace in a highly dense locality has resulted in the gradual disappearance of the historical esteros. Boncan (2015) writes:

> Informal settlers squat on the once picturesque bridges over esteros, building makeshift structures made of plywood and covered here and there with used tarpaulin. These serve not only as their pitiful places of domicile but also some form of livelihood like selling cigarettes, biscuits, candies, and cheap meals and snacks cooked from a small portable gas stove. (p. 123)

Notable tourist destinations in Binondo (Fig. 4.1) include the Binondo Church, Escolta Street, Santa Cruz Church, Chinese-Filipino Friendship Arch, Jones Bridge, and Chinatown Museum, souvenir shops, among others. More importantly, the area is frequented by both local and foreign tourists due to decades-old Chinese restaurants that serve authentic and popular cuisines such as oyster omelet and *kikiam* (e.g., Sincerity), *xiao long bao* and fried pancake (e.g., Dong Bei Dumplings), *hakaw* and white chicken (e.g., Ying Ying), *mami* (e.g., Masuki), fried *siopao* (e.g., Shanghai Fried Siopao), *buchi* (e.g., Original SaLido Restaurant), and *bihon guisado* (e.g., New Toho Food Centre) (Nuestro, 2020).[8–13]

---

[8]It is a Chinese sausage made from a mixture of shrimp and pork meat and may be served as an appetizer or a side to stir-fried dishes.

[9]It pertains to a kind of small Chinese steamed bun traditionally prepared in a *xiaolong* or a small bamboo steaming basket. It is traditionally filled with pork and is dipped in Zhenjiang vinegar.

[10]It is a noodle soup composed of wheat flour noodles, broth, and meat (chicken, beef, pork) or wonton dumplings.

[11]It is a usually thick and starchy bun fried bun filled with either pork or beef.

[12]A deep fried pastry made with glutinous rice flour coated with sesame seeds and filled with sweetened red bean paste.

[13]It is a stir-fry rice noodle dish mixed with meat, shrimp, and vegetables and is served on birthdays and other occasions.

Fig. 4.1. Tourist Attractions in Binondo, Manila. *Note:* Above (left-right): Binondo Church (by Google Maps Street View); Filipino-Chinese Friendship Arc (by Sherwin Marion T. Vardeleon); Below (left-right): Art Deco Buildings along Escolta Street and A Souvenir Shop in Binondo (by April Joy E. Dopeño).

Informality in Binondo's Tourism Value Chain: A Rapid Assessment

The informal economy is a quintessential part of Binondo's tourism industry. However, there have been a dearth in studies that delve into the former's position in the latter's value chain. Thus, the proponents utilized the tourism value chain framework developed by UNWTO (2013) with the financial support of the European Commission Directorate-General for Development and Cooperation (DEVCO). UNWTO defines tourism value chain as "a sequence of primary and support activities that are strategically fundamental for the performance of the tourism sector (p. 31)."

The framework is composed both of activities that directly and indirectly impact tourism demand. It also covers a wide array of processes such as policy-making and integrated planning, product development, promotion and marketing, distribution and sales, and destination operations and services. Further, tourism-related activities can be grouped into major categories, namely, travel organization and booking, transportation, accommodation, food and beverage, handicrafts, tourism assets in destination, leisure, excursions and tours, and support services. For this chapter, the rapid assessment only focuses on the cited activities that directly influence tourism demand. The succeeding paragraphs provide an overview of the involvement of the informal sector in Binondo's tourism industry.

Regarding transportation, both private vehicles and public utility vehicles such as buses and taxis ply the streets of Binondo. These are complemented by cheaper alternatives such as motorcycles (e.g., habal habal), tricycles (three-wheeled, motorized), and pedicabs (three-wheeled, nonmotorized), which tend to be unregistered nor covered by the law. There is no updated data on the actual number of pedicabs (Fig. 4.2) and tricycles, but the 1999 Manila City Hall records estimate that there were 10,000 total units of pedicabs in Manila, but only 1,000 were registered (Guillen & Lidasan, 2018). Guillen and Lidasan (2018) found that pedicab drivers in the city tend to borrow units from the owners, lack permits, and do not know how to drive other types of vehicles. The findings also reveal that majority of the drivers worked 5 to 7 days, earn more from May to October (rainy season), and are usually members of drivers' associations to secure protection and benefits. It may be assumed that these characteristics are shared by pedicab drivers in Binondo.

In terms of accommodation, the informal economy is less visible since most of the informal workers live in cramped spaces, makeshift houses, and have no permanent addresses. Perhaps the main area that the informal sector operates in is the food and beverage. Numerous itinerant food shops offering hot meals and staple food can be spotted around Binondo. Fig. 4.2 depicts the ubiquity of "karinderyas" or small food stalls that serve popular Filipino and Chinese cuisines at a cheaper price throughout the day. Vendors selling souvenirs and handicrafts such as key chains and magnets can also be found around Binondo.

Moreover, another well-known tourist spot in Binondo is "Divisoria" which contains stretches of stalls that offer an assortment of products for home (e.g., utensils, rugs), personal use (e.g., clothing, accessories), and other occasions (e.g., parties, reunions, holiday celebrations). Consumers can walk the streets of Tabora, Santa Elena, and Juan Luna to choose from and purchase a wide range of commodities at affordable rates. In 2020, Ylaya Street in Divisoria got a makeover when former Manila Mayor Francisco "Isko Moreno" Domagoso authorized clearing operations in the area and eventually installed around 100 newly constructed stalls equipped electric meters (Pedrajas, 2020). Vendors must likewise pay 40 pesos rental fee daily to the city government, which replaces the illegal collection system employed by some barangay officials (Pedrajas, 2020).

Binondo is also recognized as the world's oldest Chinatown that celebrates vital Chinese occasions such as the Lunar New Year and the Mid-Autumn Festival. During both events, tourist flock the area to witness cultural festivities such as fireworks show and lion and dragon dances. Several lion and dragon dance companies have been established to meet demands by Chinese individuals and businesses, and their members are engaged on a project basis (Calvelo, 2021). They thus form part of the informal economy. Concerning leisure, excursion, and tours, travel and tour agencies and independent guides have walking tours/ packages that can be availed by interested tourists and/or groups. Tour operators and guides (Fig. 4.2) may seek accreditation from the DOT.

Lastly, informal tourism can be observed in the support services. Many ambulant vendors peddle their roving "sari-sari" (micro grocery) stores to accommodate consumers who are seeking to buy a specific volume of goods, or what is called as "tingi." The sachet culture in the country is also reflected in the streets of Binondo as shown in Fig. 4.2.

In summary, the informal sector in Binondo's tourism industry is evident in the areas of transportation, food and beverage, handicraft, leisure, excursion, tours, and support services. Conversely, they are inexistent or less visible in the accommodation and tourism assets in destination segments. This is because the informal sector tends to lack ownership of properties as well as the high upkeep of maintaining rental spaces. They also prefer to engage in noncultural activities/ livelihood since these jobs ensure a steadier flow of meager income. Despite their prevalence and contributions to the cited processes, it can be posited that the informal tourism sector continues to be left out in the locality's tourism value chain and an afterthought in the local government's policy thrust and initiatives, as tackled in the next section.

Contentions and Tensions: Roadblocks Toward an Integrated Tourism Value Chain

The advancement of an integrated tourism value chain in Binondo is currently impeded by three primary, inter-related elements, namely: (i); difference in perception between decision-makers and economic actors, (ii) resistance to change by the informal sector, and (iii) lack of relevant data to support evidence-based policymaking. First, the local government unit (LGU) of Manila has traditionally viewed tourism policy planning as a state-led undertaking. As such, it has focused

Fig. 4.2. Binondo's Informal Tourism Economy. *Note:* Above (left-right): Pedicabs (by Jovito Jose P. Katigbak); Local and Foreign Tourists with a Tour Guide (by Sherwin Marion T. Vardeleon); Below (left-right): Karinderya along Mlle. de Binondo and A "Sari-Sari" Store in Benavidez Street (by Google Maps Street View).

on the implementation of DOT's national strategies such as the National Tourism Development Plan, Tourism Response and Recovery Plan, Barrier-Free Tourism, Philippine Tourism Human Capital Development Plan, and National Cruise Tourism Strategy, among others. It was only in 2022 that the DOT, Culture and Arts of Manila launched the City's inaugural Tourism and Cultural Development Plan 2020–2025, which features eight priority programs, eight trails, and eight loops located within eight tourism hubs (Manila Local Government Unit, 2022).

Moreover, Manila LGU top officials have yet to explicitly identify the informal sector's specific role in the city's tourism value chain. An illustrative case is former Manila Mayor Isko Moreno's decision to sell the Divisoria Public Market in 2020 for PHP1.4 billion to a private entity (Cayabyab, 2022). Although this was prompted by the need to effectively respond to the negative impacts of the COVID-19 pandemic, it was contrary to Mayor Moreno's objective of developing a shopping area containing vendors with permanent occupancy status (Tomacruz, 2019). Moreover, 3,700 square meter public market hosts numerous stalls that are unregistered but pay rental fees. The sale of Divisoria to a private firm means that the vendors are obliged to comply with the rules and regulations of the new owner. Notably, subsegments of the informal sector in the locality have created their respective cooperatives/associations, but this has not effectively mobilized them toward a common goal, which is to safeguard their niche in Binondo's tourism value chain. Perhaps their preoccupation with subsistence concerns, precarity of the jobs, and poor working and housing conditions have hindered them from devoting considerable resources to advancing the cited objective.

Next, the local government's efforts to formalize workers and firms have been met with resistance from the informal sector due to adverse experiences previously and the program's unclear advantages and burdensome requirements. An instructive case is the reaction of tri-wheel drivers in 2014 to the implementation of Manila City Ordinance No. 8291, which establishes the guidelines for the operation and registration of tricycles and pedicabs as well as the accompanying fees (City Council of Manila, n.d.). In particular, the tri-wheel drivers and their groups staged protests against the LGU's order banning unregistered units from traversing routes in the city's 46 main thoroughfares (Arias & Brizuela, 2014). They shared that the registration fee was an additional cost and was a way "to extort money...with these fees when we can barely earn for our families" (Arias & Brizuela, 2014). This scenario was reminiscent of the experiences of street vendors in Hanoi, Vietnam, after the government decision's to ban them due to their negative effect on the city's tourism image (Truong, 2018).

Lastly, the absence of accurate data regarding the actual contribution of the informal sector to Binondo's tourism industry hampers the formulation of empirically driven regulations. Hence, there is a need to gather relevant datasets to craft targeted policy interventions. This may be challenging for the LGU given the aloofness of the informal sector from public servants; thus, several incentives may be offered to elicit participation from the intended parties. The precarious nature of their jobs likewise requires constant data collection by the concerned authorities.

Leveraging the Informal Tourism Sector: Opportunities and Considerations

The Manila Tourism and Cultural Development Plan (MTCDP) 2020–2025 identifies Binondo, Escolta, and San Nicolas as one of the City's tourism hubs and is envisaged as the center of commerce and finance and the cradle of Filipino-Chinese heritage and culture (Intramuros Administration, 2020). Although this is a first step toward a fortified tourism value chain, several policies and initiatives may be explored to capitalize on the potential of the informal economy.

Particularly, a vital endeavor is an industry mapping of Binondo's tourism value chain led by the DOT, Culture and Arts of Manila in partnership with the relevant barangay and stakeholders such as the private sector and community-based tourism organizations. This should determine the specific role of the informal sector to formulate policies that would capacitate them to be active partners in the realization of MTCDP. The Manila City LGU may likewise collaborate with the Department of Labor and Employment and DOT in extending support and capacity-building activities to the informal workers. The Negosyo sa Kariton (NegoKart) project may serve as a model to provide livelihood assistance to tri-wheel drivers and ambulant vendors.

Further, the local government may examine the prospects of establishing linkages between formal businesses and informal enterprises to advance a more seamless value chain (Rogerson, 2018; Zhang et al., 2009). Accordingly, the LGU may erect shared service facilities specifically for the informal sector to address supply constraints and foster high productivity levels and competitiveness. It is crucial to underscore that the long-term objective should be higher volume of business registrations and workers in the formal sector. Incentives may be made available and success stories may be highlighted to elicit greater buy-in from the informal sector. Ultimately, the mainstreaming of a multistakeholder approach (Rivera & Gutierrez, 2018) is expected to build trust and confidence between and among key actors (Park et al., 2014), which may eventually result in the alignment of diverging perceptions and the formalization of the informal sector.

To address the lack of relevant data for evidence-based policymaking, the local government may then apply a regulatory sandbox approach to Binondo to test incremental yet innovative policies covering the informal tourism sector. This may substantially contribute to the development of agile regulations that are responsive to the evolving environment and needs of the cited sector. Novel initiatives may be introduced at the barangay level and subsequently expanded when they produce positive results. This approach may likewise minimize costs for the local government, which may be reallocated to the provision of various services for the informal sector. The implementation of a well-functioning monitoring and evaluation mechanism is critical in tracking the sandbox's progress and in identifying apt policy interventions.

In sum, the case of Binondo depicts that informality is historically rooted since the 66-hectare estate became the commercial and financial sector of Manila. The local government was unable to manage the continuous density growth hence resulting in cramped spaces characterized by tri-wheel drivers, ambulant vendors, roving "sari-sari" stores, and small handicraft shops. Attempts to formalize the informal sector have proven to be inadequate and oftentimes futile due to the diverging notions held by the regulator and the (un)regulated and the inherent aversion by the said sector to undergo reforms in pursuit of more sustainable outcomes. The lack of reliable and grassroots data only exacerbates the extant tensions and contentions.

Thus, it is paramount for the Manila City LGU to, first and foremost, treat the informal sector as an asset in strengthening Binondo's tourism value chain. Situating it in the locality's tourism industry is a key step in crafting accurate and effective policies and programs. More importantly, a whole-of-locality approach may be adopted wherein the primary stakeholders perform distinct yet reinforcing functions, namely: (a) local government as an agile regulator and facilitator; (b) the formal private sector as a big brother and market partner; (c) community-based tourism organizations as mobilizers, representatives, and opportunity matchmakers; (d) tourists as responsible consumers; and (e) informal sector as a capable economic agent and the little knots that bind Binondo's tourism value chain.

References

Arias, J., & Brizuela, M. B. (2014, June 24). Manila trike, pedicab drivers protest ordinance. *Inquirer.net*. https://newsinfo.inquirer.net/613858/manila-trike-pedicab-drivers-protest-ordinance#ixzz8Os01gI5j

Back, L. (2019). *Informal economy in the context of globalization and urban gentrification – The case of small-scale farmer-vendors in the City of Naga, Philippines* (Publication No. 169838) [Master Thesis, Stockholm University]. Stockholm University Publications.

Boncan, C. P. (2015). Binondo yesteryears today: Continuity and change. In A. T. Torres, L. L. Samson, & M. P. Diaz (Eds.), *Filipino generations in a changing landscape* (pp. 105–129). Philippine Social Science Council.

Cakmak, E., Lie, R., & McCabe, S. (2018). Reframing informal tourism entrepreneurial practices: Capital and field relations structuring the informal tourism economy of Chiang Mai. *Annals of Tourism Research*, *72*, 37–47. https://doi.org/10.1016/j.annals.2018.06.003

Calvelo, G. (2021, February 12). Binondo dragon dancers welcome Lunar New Year minus the bang. *ABS-CBN News*. https://news.abs-cbn.com/life/multimedia/slideshow/02/12/21/binondo-dragon-dancers-welcome-lunar-new-year-minus-the-bang

Carlos, J. C. T., Gutierrez, E. L. M., & Bautista, M. J. D. (2023). *Gendering the informal tourism sector toward inclusive and sustainable growth: The case study of Boracay Island*. PIDS Discussion Paper Series No. 2023 35. https://pidswebs.pids.gov.ph/CDN/document/pidsdps2335.pdf

Cayabyab, M. J. (2022, March 29). Moreno defends sale of Divisoria public market. *PhilStar Global*. https://www.philstar.com/headlines/2022/03/29/2170566/moreno-defends-sale-divisoria-public-market

Caynila, K. A., Luna, K. T., & Milla, S. A. A. (2022). *The Philippine tourism sector amid the pandemic: Developments and prospects*. Bangko Sentral ng Pilipinas

Economic Newsletter No. 22-02. https://www.bsp.gov.ph/Media_And_Research/Publications/EN22-02.pdf

Chen, M. (2016). *Expanding the economic potential of women informal workers*. UN Secretary-General's High-Level Panel on Women's Economic Empowerment. http://www.wiego.org/sites/default/files/resources/files/WIEGO-expanding-econ-potential-informal-workers.pdf

Chongvilaivan, A., & Chooi, A. (2021). *A comprehensive assessment of tax capacity in Southeast Asia*. Asian Development Bank. http://dx.doi.org/10.22617/TCS210243

City Council of Manila. (n.d.). *Ordinance No. 8291*. City Ordinances. https://citycouncilofmanila.com.ph/city_ordinances/ordinance-no-8291-2/

Department of Tourism (DOT). (2023a). PTSA derived indicators 2000–2022. http://tourism.gov.ph/files/2023/07/07-17/PTSA%20Derived%20Indicators,%202000-2022.pdf

DOT. (2023b). DOT chief unveils National Tourism Development Plan (NTDP) 2023–2028 at stakeholders' summit. [Press release]. https://beta.tourism.gov.ph/news_and_updates/dot-chief-unveils-national-tourism-development-plan-ntdp-2023-2028-at-stakeholders-summit/

Development and Cooperation Europeaid (DEVCO), & United Nations World Tourism Organization (UNWTO). (2013). *Sustainable tourism for development guidebook*. https://doi.org/10.18111/9789284415496

Dreyer, M., & Nygaard, K. (2020, May 13). The Philippines provides support to workers in the informal economy. *Program on Financial Stability*. https://som.yale.edu/blog/the-philippines-provides-support-to-workers-in-the-informal-economy

Gonzales, M. L. J. (2018). *Size of the informal economy in the Philippines*. International Labor Organization. https://www.ilo.org/wcmsp5/groups/public/—asia/—ro-bangkok/—ilo-manila/documents/presentation/wcms_634914.pdf

Gonzales, M. L. J. (2022). *National employment recovery strategy for the workers in the informal sector*. Resilience and Innovation Learning Hub. https://rilhub.org/wp-content/uploads/2022/02/Good-Practices-in-Social-Protection-in-the-Informal-Economy-04-NERS-for-the-Informal-Sector.pdf

Guillen, M. D. V., & Lidasan, H. S. (2018). *A study of the non-motorized public transportation in urban and urbanizing areas: The case of pedicab operations in the city of Manila and in the Municipality of Los Banos, Laguna*. National Center for Transportation Studies University of the Philippines Diliman. https://www.ncts.upd.edu.ph/tssp/wp-content/uploads/2018/08/Guillen00.pdf

International Labor Organization (ILO). (2002). *Resolution concerning decent work and the informal economy*. International Labor Organization. https://www.ilo.org/public/english/standards/relm/ilc/ilc90/pdf/pr-25res.pdf

ILO. (2018, October 10–19). Resolution concerning statistics on work relationships [Conference session]. In *20th International Conference of Labour Statisticians*, Geneva, Switzerland. https://www.ilo.org/wcmsp5/groups/public/—dgreports/—stat/documents/meetingdocument/wcms_648693.pdf

ILO. (2023a). *Women and men in the informal economy: A statistical update*. International Labor Organization. https://www.ilo.org/wcmsp5/groups/public/—ed_protect/—protrav/—travail/documents/publication/wcms_869188.pdf

ILO. (2023b, October 23). *New standards, increased visibility: Improving measurement of the informal economy*. [Press release]. https://ilostat.ilo.org/new-standards-increased-visibility-improving-measurement-of-the-informal-economy/

Intramuros Administration. (2020, December 13). *Manila's tourism and cultural development plan.* [Press release]. https://intramuros.gov.ph/2020/12/13/manilas-tourism-and-cultural-development-plan/

Lv, Z. (2020). Does tourism affect the informal sector? *Annals of Tourism Research, 80*, 102816. https://doi.org/10.1016/j.annals.2019.102816

Malogdos, F. K., & Yujuico, E. (2015). Reconciling formal and informal decision-making on ecotourist infrastructure in Sagada, Philippines. *Journal of Sustainable Tourism, 23*(10), 1482–1503. https://doi.org/10.1080/09669582.2015.1049608

Manila Local Government Unit. (2022). Unveiling of the Manila tourism and cultural development plan 2020–2025 and tourism stakeholders meeting. [Press release]. https://www.cityofmanila.ph/articles/647

Milgram, B. L. (2014). Remapping the edge: Informality and legality in the Harrison Road Night Market, Baguio City, Philippines. *City and Society, 26*(2), 153–174. https://doi.org/10.1111/ciso.12038

National Economic Development Authority. (n.d.). *Philippine development plan (PDP) 2023–2028.* https://pdp.neda.gov.ph/philippine-development-plan-2023-2028/

Nuestro, E. (2020, January 20). Binondo food trip 2020: Where to get the best Chinese delicacies in Manila's old Chinatown. *Tatler Asia.* https://www.tatlerasia.com/dining/food/a-gastronomic-adventure-in-binondo

One Planet Network. (n.d.). *Overview and hotspots analysis of the tourism value chain in the Philippines.* https://www.oneplanetnetwork.org/sites/default/files/iki_country_report_philippines_final_1.pdf

Park, J., Lee, H., & Kim, C. (2014). Corporate social responsibilities, consumer trust and corporate reputation: South Korean consumers' perspectives. *Journal of Business Research, 67*(3), 295–302. https://doi.org/10.1016/j.jbusres.2013.05.016

Pedrajas, J. (2020, August 22). Ylaya street, Divisoria's 'textile capital' gets a new look. *Manila Bulletin.* https://mb.com.ph/2020/08/22/ylaya-street-divisorias-textile-capital-gets-a-new-look/

Philippine Statistics Authority. (2023, January 18). *2020 survey of tourism establishments in the Philippines (STEP) – Economy-wide: Preliminary results.* https://psa.gov.ph/content/2020-survey-tourism-establishments-philippines-step-economy-wide-preliminary-results

Presidential Communications Office. (2023). PBBM approves five-year national tourism development plan. [Press release]. https://pco.gov.ph/news_releases/pbbm-approves-five-year-national-tourism-development-plan/

Rivera, J. P. R., & Gutierrez, E. L. M. (2018). A framework toward sustainable ecotourism value chain in the Philippines. *Journal of Quality Assurance in Hospitality & Tourism, 20*(2), 123–142. https://doi.org/10.1080/1528008X.2018.1492495

Rogerson, C. M. (2018). Informal sector city tourism: Cross-border shoppers in Johannesburg. *GeoJournal of Tourism and Geosites, 22*(2), 381–392. https://doi.org/10.30892/gtg.22209-296

Royeca, G. (2023, February 28). Informal power. *The Philippine Star.* https://www.philstar.com/opinion/2023/02/28/2248080/informal-power

Shead, B. (2017, July 26). Tourism industry in the Philippines: Part I. *ASEAN Briefing.* https://www.aseanbriefing.com/news/tourism-industry-philippines-part/

Slocum, S. L., Backman, K. F., & Robinson, K. L. (2011). Tourism pathways to prosperity: Perspectives on the informal economy in Tanzania. *Tourism Analysis*, *16*(1), 43–55. https://doi.org/10.3727/108354211X12988225900045

Tomacruz, S. (2019, July 9). Isko Moreno's clearing operations: Which Manila areas have been covered? *Rappler*. https://www.rappler.com/nation/234975-list-isko-moreno-clearing-operations-manila/

Truong, V. D. (2018). Tourism, poverty alleviation, and the informal economy: The street vendors of Hanoi, Vietnam. *Tourism Recreation Research*, *43*(1), 52–67. https://doi.org/10.1080/02508281.2017.1370568

UNWTO. (2021, January 28). *2020: Worst year in tourism history with 1 billion fewer international arrivals*. News. https://www.unwto.org/news/2020-worst-year-in-tourism-history-with-1-billion-fewer-international-arrivals

Williams, A. M., & Shaw, G. (2011). Internationalization and innovation in tourism. *Annals of Tourism Research*, *38*(1), 27–51. https://doi.org/10.1016/j.annals.2010.09.006

Zhang, X., Song, H., & Huang, G. Q. (2009). Tourism supply chain management: A new research agenda. *Tourism Management*, *30*(3), 345–358. https://doi.org/10.1016/j.tourman.2008.12.010

II

"Thinking Small" as a New Pillar of Sustainable Tourism

Chapter 5

Going Small Yet Getting Big: Paradigm Shifts in Business Models Toward Inclusivity, Sustainability, and Long-Term Resilience and Proofing in Micro and Local Tourism

Luisito C. Abueg, Ma. Janda Ira Felina M. Benedictos and Claire Therese B. Villafuerte

University of the Philippines Los Baños, Philippines

Abstract

The COVID-19 pandemic has reduced much of the complexities of human life, which has affected the multifaceted socioeconomic dimensions and sectors around the world. One of the main consequences of the pandemic is the reduction in physical mobility and shifting toward a more digitally enhanced platform, where activities may be continued given the imposed physical limitations. Global supply chains were broken down into regional and even local ones and travel has been reduced significantly across leisure and nonleisure reasons, among others. While we recognize that these changes would require substantial capital and financial investments, these will not be put to waste given that emerging models, practices, and trends are the "silver linings" of the pandemic onto the tourism sector. Given that one emerging advocacy is the sustainability of tourism destinations under the backdrop of environmental awareness, care, and sustainable use, we also look at how tourism goods and services are put on sale given the adoption of new models. Finally, we envelope the pandemic experience and the induced iterations of local tourism initiatives as an opportunity to look for new and revitalized tourism innovations toward the "next normal."

Revisiting Sustainable Tourism in the Philippines, 87–106

Published under exclusive licence by Emerald Publishing Limited
doi:10.1108/978-1-83753-678-820241005

Keywords: Microenterprises; COVID-19 pandemic; local tourism; local economic growth; economic resilience

Introduction: The Business Climate and the COVID-19 Pandemic

At the onset of the pandemic in early 2020, quarantine and travel restrictions were imposed to minimize the spread of the novel coronavirus. This forced people to remain in their residences, driving economic activities to a slowdown all over the world. In the Philippines, the Department of Trade and Industry (DTI) recorded a decline of almost 43,000 establishments between 2019 and 2020, most of which are from micro- and small-enterprises (DTI, 2019, 2020). Micro-, small-, and medium-scale enterprises (MSMEs) are considered the backbone of the economy as these enterprises comprise 99% of the total number of businesses in the country and contribute more than 60% of the total employment, to date. MSMEs were largely affected by the pandemic and are more vulnerable to shocks due to their limited resources (Shinozaki & Rao, 2021). From this survey, more than 70% of the surveyed enterprises temporarily halted operations due to the pandemic (Shinozaki & Rao, 2021). The mobility restrictions also caused delays in the delivery of products and services and disruption of production and supply chain for the firms. In the same survey, several other concerns were also reported such as the decline in domestic and foreign demand and the cancellation of contracts which amplified the decline in the operations of the enterprises. A disaggregation of the results by sector shows that 70% of the enterprises from the service sector temporarily closed. Both the manufacturing and agriculture sectors show that a significant number of establishments temporarily closed.

The challenges faced by businesses during the pandemic translated into 8.7% unemployment rate and 14.4% underemployment in October 2020, a 4.2% and 1.4% increase, respectively, from the reported October 2019 data of the Philippine Statistics Authority (PSA) during the pandemic years (PSA, 2020, 2021). These struggles were not unique to the Philippines and were also observed in other economies. In neighboring countries, results from a similar survey showed that around 48.6% in Indonesia, 61.1% in Lao PDR, and 41.1% in Thailand temporarily closed business due to the pandemic (ADB, 2020a). Businesses that continued operations also saw a decline in their domestic and foreign demand.

To revive economic activities, lockdowns were slowly eased in areas with lower cases. In the Philippines, legislations such as Bayanihan 1 (Republic Act [RA] no. 11469) and Bayanihan 2 (RA no. 11494) were also signed which includes provisions for financial aid to help businesses recover and to stimulate household consumption (Abueg, 2020a). In adapting to the demands of the "new normal," businesses were compelled to overhaul their operational strategies. These practices include finding local sources of input, digitization, moving functions that can operate remotely to work-from-home setup, and improving their technological infrastructures. For example, MSMEs in the Davao region in the Philippines were considering upgrading their technology, particularly in e-commerce, factory operations, and automation (Laorden et al., 2021). The pandemic also gave way to the rise of digital payments, according to World Bank Group (2022), with 57% of those who used digital payments in 2021 being first-time users after the COVID-19 pandemic started, in which firms need to acclimate to.

In this chapter, we look at the activities and models in the tourism sector before and during the pandemic. We look at these and trace the developments as the sector adjusts, calibrates, and realigns itself to the constraints and conditions of the "new normal." These new forms of activities and modes of businesses have also made the tourism sector rethink the prepandemic models and strategies to keep businesses afloat, surviving, and be prepared for similar shocks in the future. Apart from additional health compliance checks, new tourism models have downsized into smaller units and scopes: the so-called "micro-tourism" models. We look at these new models together with the added digitalization elements of local tourism to find ways and means of improving the business climate despite the stresses of the pandemic and the requirements toward postpandemic viability. We look at new forms of marketing tourism goods and services, as well as new specifics of destination tourism, and see how these new iterations affect prepandemic human capital investments in labor, training, and specialized work.

For example, the digitization requirements allowed more data-driven decision-making across all stakeholders and layers of decision-making, both from the public sector and private enterprises. We also see the continued relevance of moral suasion, advocacy, and recognition of the importance of supporting local tourism and the creation of "common goods," but such works only with the required support of the government and other stakeholders. Apart from providing updated and enhanced guidelines, health certifications, and safety seals, there must be a revisit of the overarching national program and alignments of respective local tourism plans to make the renewed experience of local and international tourists seamless, enticing, and would encourage repeated behaviors. Toward the end of this chapter, we argue that these small-scale movements, models, and iterations are arguably to be maximizing potentials as the tourism sector transcends to the postpandemic situation, making these small-scale movements big time or maximal in terms of benefits to stakeholders and the economy.

Micromovements in Tourism

In the Philippines, the tourism industry was and still is greatly affected by the COVID-19 pandemic. The pandemic has resulted in a decrease in tourist arrivals in the country, which in turn has affected employment in the tourism sector, in terms of job generation and the sector's absorption of the labor force (ILO, 2021; PSA, 2022). It is fortunate that given the eased restrictions and improving health conditions beginning 2021, the sector has shown signs of recovery and potential growth that are at par or may be beyond the prepandemic data. Data figures in this chapter indicate such silver linings (Table 5.1). It shows the comparison from 2019 to 2022, given the tourism indicators of the United Nations World Tourism Organization (UNWTO).

Table 5.1. Tourism Indicators for the Philippines, 2019–2022.

Indicator	2019	2020	2021	2022
International tourist arrivals (in million)	8.3	1.5	0.2	2.7
International tourism receipts (in USD billion)	9.8	1.8	0.6	4.2
International tourism expenditure (in USD billion)	12	4.1	3.3	4.9
Tourism as share of total exports (in percent)	10.3	2.6	0.9	13.2
Tourism as share of GDP (in percent)	8.6	8.6	8.6	8.6
Employment in tourism (in million)	5.7	4.7	4.9	No data
Beds per 1,000 population	2.3	3.1	3.8	No data
Share of top 3 months in annual arrivals (in percent)	27	94	38	92

Source: UNWTO.

The pandemic has greatly affected the share of tourism as the country's exports. Arrivals have decreased by almost 82% by 2020 and have no sign of increasing by 2021 (Table 5.1). With the easing of the travel restrictions by 2022, the country has seen an increase in the arrivals of tourists in the country but is still far behind the prepandemic numbers. This calls for strategies that could make tourism in the Philippines sustainable and maintain its competitiveness. These are emboldened in Chapter 7 (p. 179) of the Philippine Development Plan (PDP) 2023–2028 (NEDA, 2023).

The Organization for Economic Co-operation and Development (OECD, 2020) has identified key policies by the end of 2020, which should be prioritized to minimize job losses and rebuild recovery in 2021. The following were the key agreements:

- Restoring traveler confidence.
- Supporting tourism businesses to adapt and survive.
- Promoting domestic tourism and supporting the safe return of international tourism.
- Providing clear information to travelers and businesses and limiting uncertainty (to the extent possible).
- Evolving response measures to maintain capacity in the sector and address gaps in supports.
- Strengthening co-operation within and between countries.
- Building more resilient, sustainable tourism.

With these issues, this chapter looks at micro-tourism, with the abovementioned policies from the OECD. The discussions were heavily centered on (a) promoting domestic tourism and (b) building more resilient and sustainable tourism.

Clearly, the Philippines, domestic tourism drives tourism behavior before and after the pandemic. As defined by the UNWTO, domestic tourists are locals who travel within the country and stay overnight. They also have a category of domestic visitors, but for the Philippines, unfortunately, no data indications for such metric. This highlights the importance of looking into the numbers pertaining to domestic tourism led by counts of domestic visitors. In a report by the United Nations Conference on Trade and Development (UNCTAD) in 2020, as restrictions were lifted, people are now returning to travel but still have fears of traveling internationally. Thus, these perceptions create profound effects on economies that are dependent on tourism, which the Philippines is doing in recent decades. UNCTAD (2021) concluded that governments should diversify and not only rely on one sector; there is a need to capacitate its citizens through education and training programs that could boost other targeted economic sectors. In the Philippines, apart from overseas remittances that started in the 1970s, business process outsourcing became the next sector to yield employment and value-added for the macroeconomy and now the tourism sector (Dacanay et al., 2018).

Firstly, the case of the Philippines in terms of its tourism sector has shown great dependence on international tourists, whose numbers dwindled when the pandemic hit. While it is evidenced that the number of domestic tourists in the country indeed from 2011 to 2019 before the pandemic experienced a slow but steady growth, the pandemic clearly wiped out the momentum of the sector in terms of value-added and contribution to the national product (Fig. 5.1).

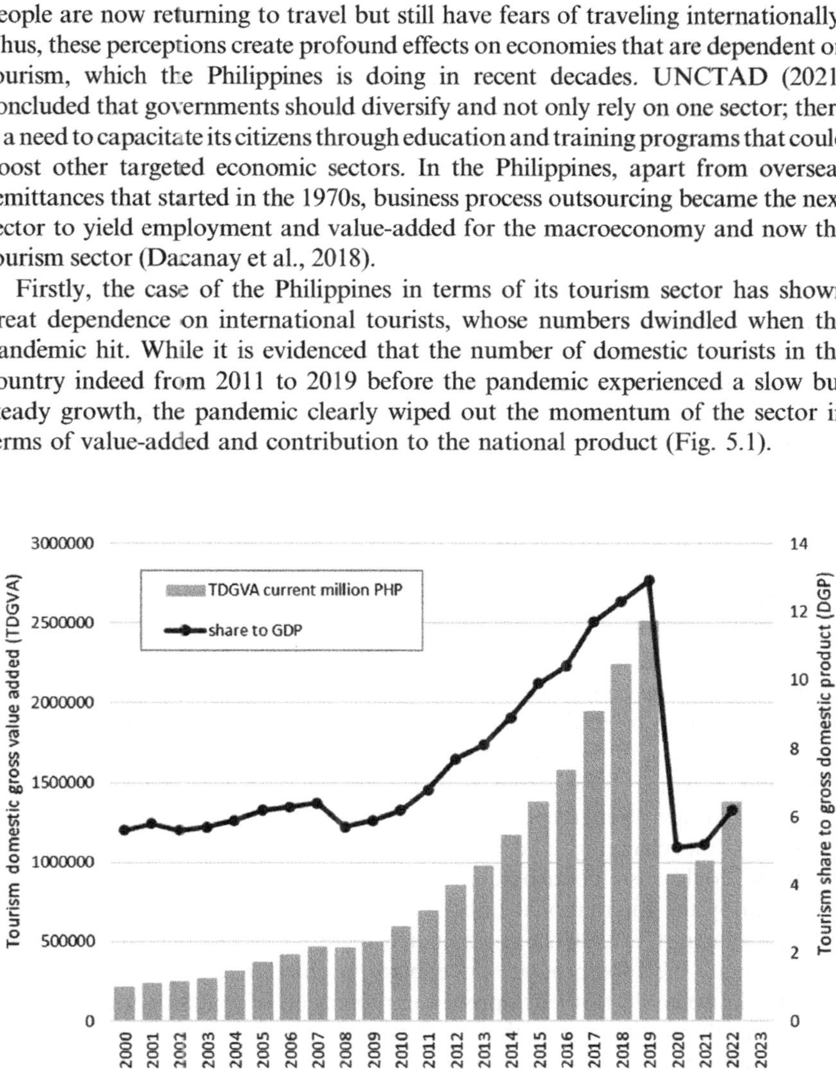

Fig. 5.1. Tourism Domestic Gross Value Added (in Current Million PHP, Left Axis) and Share to Gross National Product (in Percent, Right Axis), 2000–2022. *Source:* PSA.

Note also of the peak in 2019, which could be explained by the opening of Boracay in the last quarter of 2018 after its much-needed rehabilitation and the rehabilitation of the commercial airports in the country (Fig. 5.1). Former Department of Transportation (DOTr) Secretary Arthur P. Tugade reported that 17 commercial airports have been rehabilitated in 3 years and 28 are ongoing (Marasigan, 2019). In addition, it was also in the same year that the Department of Tourism (DOT) launched the slogan "#MoreFunForever," a campaign which was reported to have the three-pronged campaign. The program aims to (a) promote responsible tourism, (b) enforce environmental compliance, and (c) improve the quality of service offered by stakeholders while supporting local communities (Japan Times, 2019). Looking at the aims of this campaign of DOT, this slogan and program can greatly support the micro-tourism model.

Domestic tourism has been a bigger contributor relative to inbound tourism in the last 23 years. The slogan making of DOT and promotion of local tourism portfolio – destination, travel and leisure experience, gastronomy, local culture, goods and services – may be effective to drive demand for nationals to travel to various destinations in the country. Zhao (2022) defines micro-tourism as "a short-term improvisational leisure experience, where tourists use internet platforms to choose nearby locations where they could spend their leisure time." This new trend is essential when we look at the first steps of the tourism sector in its way to recovery postpandemic. More particularly, as local tourism is driven by "revenge spending" and "revenge travel" (Dy-Zulueta, 2023), there are also instances that rural destinations benefit from such phenomenon (Panzer-Krause, 2022). However, it must be noted with caution that these phenomena were driven by pandemic fatigue and mobility restrictions, thereby an in-depth look at these emerging movements must be looked at, considering stakeholders and the tourism sector at large (Wang & Xia, 2021; Zaman et al., 2021).

Secondly, for micro-tourism to flourish, MSMEs should be also incorporated in the holistic approach of the program, as these enterprises also form part of the tourism sector. The cooperation between private and public institutions needs to be strengthened to boost domestic tourism in the country. In 2018, 63.19% of the employed members of the workforce were from MSMEs. One of the major issues is funding, in which the government must implement policies and programs that are geared toward strengthening these enterprises (Raquiza, 2021).

Thirdly, the changing landscape in tourism before, during, and after the pandemic could also contribute in boosting domestic tourism. In a toolkit developed by the International Labor Organisation (ILO) on poverty reduction through tourism (ILO, 2012), it was noted there the elements of the marketing mix which are the 5 Ps: product, price, place, promotion, and people. Using this framework, we could further improve the promotion of domestic tourism.

In terms of product, the destination acts as the portfolio of tourism goods and services. For example, the expenditure of travelers in terms of accommodation, entrance fees, transportation fees, food, souvenir items, and other expenditures related to this travel will affect tourism spending. This is where MSMEs, firms, local citizens, and the local government unit (LGU) come into the picture. In a report by the Canadian Trade and Investment Facility (CITF) for Development on the post-COVID-19 recovery plan for ASEAN tourism, it proposed that MSMEs should be given support for them to become resilient in the future (CITF, 2021). By strengthening MSMEs, we will be able to create a competitive environment, which may help reduce prices. To do this, LGUs, local citizens, and firms should cooperate and coordinate among each other, so that effective strategies be formulated.

In terms of the place or destination, technology and digitization in tourism are not only limited to the physical space. In an article by Lee et al. (2020), they have stated the rise of smart cities and smart tourism. Augmented reality (AR) and virtual reality (VR) may be used as part of the steps to make tourism take advantage of impending uncertainties brought forth by quarantines, lockdowns, and restrictive health protocols that may be still looming in the corner. The place or destination in tourism enterprises is not the only thing that is affected by the recognition of the Fourth Industrial Revolution. Lee et al. (2020) furthered that with the information communication technology (ICT) infrastructure, it will be much easier to plan, preevaluate, and postevaluate the experience of tourists and travelers. The face and methods of promotion in tourism have also changed and are continuously evolving, given the events before and after the pandemic. This is why the people factor (the last in the 5Ps of the ILO framework) is equally essential. Buonincontri and Micera (2016) argued, for example, that in evaluating the smart tourism frameworks of the cities of Venice, Italy, and Salzburg, Austria, they noted that the cocreation of tourism experience is an integral part for people to share their experiences.

In terms of technology, the Asian Development Bank noted that the COVID-19 pandemic has somewhat paved the way for firms to digitalize; they highlighted that countries like Fiji, Papua New Guinea, and Samoa have included in their nation development policy the digitalization of their economies. Additionally, MSMEs are greatly involved in the industry as they are actively participating in the activities of the business sector (ADB, 2023). Indeed, technology has changed the way that tourism is being promoted, and it also strengthens the MSMEs with the right policies and programs in the digitalization of the economy.

However, the pandemic has unraveled a lot of “hidden” weaknesses and gaps in the tourism industry. One of the pressing issues that we must incorporate as necessary is the observance of health protocols that should be implemented when tourists enter our destinations. Initial guidelines of the government on mobility might be found to be restrictive by the sector’s stakeholders. Additionally, capacity-building trainings for local citizens in the destinations and the sector’s stakeholders must be provided to make sure that if this similar crisis happens again, the effects would be mitigated.

Notably, in the country’s PDP 2011–2016, a vision of having a “safe, secure, efficient, viable, competitive, dependable, integrated, environmentally sustainable and people-oriented Philippine transport system” is indicated and one of its major themes. A subsequent Technical Education and Skills Development Authority (TESDA) (2017) report in relation to the PDP noted that the maritime industry is vital in the socioeconomic progress of the country. In addition, they have also pointed out the possibility of partnering with DOT in terms of pushing for the cruise industry in the country which at present is being bolstered by the DOT. Given that the tourism industry in the Philippines (as indicated in Fig. 5.1) is recovering, it is with utmost importance that the country also turns its eye toward developing further and updating the skill sets and abilities of our seafarers, as they become competitive in the global arena.

The world is recovering from the aftereffects of the COVID-19 pandemic, and this is also an important opportunity for the Philippine tourism industry – together with its stakeholders – to rise from the challenges that it has posed. This is clearly highlighted with the data indications and programs in the 2023–2028 PDP, which envisions a revival, bounce back, and even heightened growth of tourism. Note that the sector remained to as a source of value-added for the Philippine economy together with overseas remittances and business process outsourcing (Abueg, 2023b; Caynila et al., 2022; Dacanay et al., 2018). With these arguments, the future of the sector and its stakeholders faces a significant crossroad: the continued relative advantage of the sector amid the increasing competition in the global arena. As such, data indications from the Bangko Sentral ng Pilipinas (BSP) show that sea-based remittances are continuously increasing in terms of share to total remittances. These indications may be attributed to most seafarers and maritime workers contributing to sea-based remittances, of which tourism is a part of (Acejo & Abueg, 2022).

Business and Economic Adjustments to the "New Normal"

One of the glaring facets of the COVID-19 pandemic is that health standards are becoming one important facet in everyday life, including business models and transactions. With the tourism sector moving forward, health and safety protocols and standards are put in place, with monitoring, evaluation, and policy iterations, thanks to the accelerated technological migration. Such acceleration being one trend even before the pandemic becomes pressingly important given the effects and the recovery efforts from the pandemic (Abueg, 2020b). This recognition has been the theme of the World Tourism and Travel Council in launching of its "Rediscovering Tourism" in 2022 (Tayag, 2022).

As in the usual cliché in economics, "there is no such thing as free lunch." Not free, because not only financial costs are required in these new iterations, but also recalibrations of all prepandemic routines. These recalibrations start from revisiting all protocols, use of inputs, practices, and other related business-related functions. This is also true in tourism since the pandemic is not a simple virus that is wreaking havoc in human life, but rather creating significant and permanent changes in various aspects of social affairs. We start with the proper and adequate information on the pandemic events – thanks and a curse attributed to technology. Thanks, because information becomes readily available, and we can recalibrate our methods in shorter time spans. Curse, because with digitization, there is another "dark side" of the pandemic: the triumvirate of mis-, dis-, and mal-information. Thus, it is important that health standards are backed with proper data to yield sound decision-making for policymakers.

Next, we also highlight the seemingly permanent changes in business entities as we navigate toward the postpandemic era. Businesses become more hybrid, having both online and onsite counterparts. There is also some "democratization of businesses" and the creation of new business opportunities through the demand-inducing smartphone technology aided with internet technology (Ganti, 2020), from the old shopping malls regarded as "cathedrals of consumption" (Rico & de Leon, 2017). We also look at this phenomenon as indicated by other forms of entertainment media (Abueg, 2021) and in the way society struggles with survival, especially during the pandemic (Abueg, 2023a). A particular case in point is that the country's fiscal managers saw the need to revive decongestion problems of urban centers in the country, but it again reiterated that such a program must be coupled with sustainable economic opportunities in recipient locations so as not to fall prey into vicious cycles and pitfalls of sound urban planning and economic programming (Correa & Abueg, 2020).

We investigate these developments and how the tourism sector can adapt to the pressing changes and incorporate innovations in the most cost-efficient and least disruptive manner. On the one hand, we see that technology has created opportunities for the tourism sector to continue to offer their goods remotely via online platforms. For example, a lot of digital outlets were created for goods that have sustained demand, e.g., those from Baguio City, Cebu City, and Davao City. These digital platforms have also aided in tourism promotion, which led to the notion of "revenge spending" (Abueg, 2020b), and later, the notion of "revenge travel" (Marcus, 2022). On the other hand, technology adoption entailed costs of acquisition, training of existing and new personnel, and the required alignments necessary for the postpandemic operations of tourism businesses and stakeholders.

Even before the pandemic, data from PSA reveal that the tourism sector is dominated by MSMEs, also called microenterprises. These microenterprises are recognized by the national government as a major component for realizing future-proofing plans of the sector, moving forward from the pandemic, and are expected to perform better in the next years (Arnaldo, 2020; DOT, 2022). Further, the importance and integral role of microenterprises in the tourism value chain warrants the intended support to bolster the potentials of the sector (Padronia, 2023).

Notably, it is being recognized that even before the pandemic, microenterprises generally contributed more to total employment and are considered to be the drivers of value-added across sectors in the economy (World Bank Group, 2019). Arnaldo (2020) furthered that the prepandemic data from PSA suggest close to a 100% of establishments in the tourism sector are microenterprises. While there is a relatively small decline in employment despite increased number of establishments in 2020 (PSA, 2023), ILO (2021) noted that the Philippines is among the five countries with the largest job losses that have occurred due to the pandemic. Nevertheless, the recent 2021 national data on establishments and employment showed that microenterprises still have a large share in employment, with tourism as one of the sectors contributing significantly to employment (PSA, 2022). The recovery of the sector from the years of the pandemic manifested in recent MSME data, showing that the sector is back on track with employment numbers and its share of total employment (DTI, 2022).

On a side note, the COVID-19 pandemic also served as a "breather" to tourist destinations that capitalize on natural resources and environmental elements. These manifestations of "environmental healing" also provided insights on renewed environmental sustainability on ecotourism. This is also amplified given the developments happening in climate change patterns, which are seen to affect the environmental capital of tourist destinations. Such attention is also linked by scholars to the increasing affluence of people and the consumerist public (Wiedmann et al., 2020).

Apart from environmental impacts and implications of recalibrating business models on tourism, there is also a renewed call for supporting the local economy as the country starts to open postpandemic. Tourism authorities also advocated for supporting local tourism by encouraging local tourism travel and providing tourism sector stakeholders with incentives and support. This program is not new and has undergone several changes in previous decades (e.g., Abueg, 2005). However, given that even in the prepandemic years tourism has been recognized to contribute significantly to economic value-added (Abueg, 2023b), the COVID-19 pandemic created a renewed call for support of local industries, particularly on tourism. As early as 2021, the gradual reopening of the economy with eased restrictions and recalibrated tourism programs (with health guidelines) contributed to the remarkable increase in local tourism activity noted by the national government (Rocamora, 2021). This is even one of the continued programs with the change of national leadership and more stakeholder participation (Gutierrez, 2021).

While it is good and remarkable that there is a consistent call to support the local economy in general and local tourism in particular, the business sector continues to innovate existing models incorporating such advocacy (Abueg et al., 2014). However, this premise is necessary but not sufficient, in the sense that there must be a complementary redefinition of perspectives in business through pro-social models taught in business schools, both in formal and informal modes (Abueg et al., 2016). Note that these interventions of moral suasion may not be sustainable, since the problem may be casted in economic language as the classical principal–agent problem. This class of problems results in the known free rider (a particular case of the moral hazard) problem. Bowles (2016, p. 2) argued that this may be contributed by economic conditions and what it favors:

> If I am right, then an erosion of the ethical and other social motivations essential to good government could be an unintended cultural consequence of policies that economists have favored, including more extensive and better-defined private property rights, enhanced market competition, and the greater use of monetary incentives to guide individual behavior.

While a lot of civic society initiatives worked on the community spirit through community pantries (Abueg, 2023a), it is commendable that internal support among affected stakeholders, businesses, and affected workers manifested through these initiatives (e.g., in Ocampo, 2021). These private initiatives were even supported and welcomed by the national government, recognizing the need for collective support for the sector with aid not only coming from the public sector but also to those individuals in their private capacity (Adel, 2021). Apart from community pantries, another important nonmonetary-based form of trade and transactions are online barter trades. These technology-inspired crude forms of transactions re-emerged during the COVID-19 pandemic. Remarkably, technology infusion mitigated the problem of double coincidence of wants, an economic problem which is the feature of the early forms of barter (Abueg, 2023a). In tourism-affected Siargao island, it claims to have started the first online barter. This online barter platform also targets to help affected tourism workers in the island, days after the declaration of the lockdown in March 2020 (Catoto, 2020). Also, as pointed out in the case of community pantries and online barter, proper identification of issues and problems and a multistakeholder approach to collective planning must be the platform moving forward (Abueg, 2023a). More importantly, the continuously evolving multistakeholder nature of tourism must always be recognized at the forefront of research, policy and governance, and business modeling, among others (Farsari, 2021).

Both research literature and anecdotal evidence point to these classical problems of the concept of goodness being necessary but not sufficient in promoting sustainability: workers and microenterprises being resilient or tourists and travelers supporting tourism goods and services with the branding of advocacy (Abueg et al., 2014). As in the case of tourism, these nuances are important to be recognized but should receive ample support both from the public sector and from civil society to attain sustainability and to realize long-term benefits. All of these, being at the backdrop of the postpandemic "next normal," some of the proposed interventions are provided for in the next section.

Public Enterprises, Government Support, and Program and Policy Adjustments

To ensure a seamless transition to reopening economic activity, a joint memorandum circular (JMC) was issued in 2021 collaboratively by the following government agencies:

- Department of Labor and Employment (DOLE).
- Department of the Interior and Local Government (DILG).
- Department of Health (DOH).
- DTI.
- DOT.

These agencies have agreed to issue "safety seals" to their respective stakeholders under their jurisdictions in the conduct of business. These businesses are provided certifications of compliance with the set minimum public health standards by the government (DILG, 2021). For DOT, it encourages stakeholders to support these entities with issued safety seals (Rocamora, 2022). For the tourism establishments, DOT is tasked to issue safety seals for the following (as stipulated in the JMC):

- Primary tourism enterprises: accommodation establishments (hotels, resorts, apartment hotels, and other accredited accommodation establishments); travel and tour services; meetings, incentives, conferences, and events (MICE) venues and facilities.
- Restaurants inside hotels and resorts.[1]

It is also a noteworthy coincidence that one important legislation prepandemic that became one distinct feature of the pandemic months was the work-from-home (WFH) arrangements. This was anticipated earlier in 2018 by the Philippine Legislature, passing Republic Act no. 11165 (Official Gazette of the Philippines, 2018). Although the justification was mainly due to the worsening traffic conditions in urban centers, this was essentially the key setup during the COVID-19 pandemic. Some businesses have continued to do a mix of WFH and onsite work, called hybrid work arrangements. WFH arrangements also supported another prepandemic notion of work, called "workcation." This is a combination of work and leisure, usually working remotely but spending time as well in a tourist destination or some related tourist accommodation (Pecsek, 2018).

Both the idea of WFH and workcation may only be known to some during the COVID-19 pandemic, but these ideas became so popular that they gave another opportunity for the tourism sector to nudge some areas for innovation and improvement. Breaking from the traditional dichotomy of work and leisure, these practices are promoted in many parts of the world. These not only benefit the tourism sector but also promote better human well-being and productivity in the workplace. Some tourist visits are mixing workcation with revenge travels, which ultimately benefits the tourism sector.

The national and local governments are also equally active in promoting tourism and going back to the target of increasing tourism portfolio and value-added for the economy. With these, the Philippine government has adopted one policy of rebating value-added tax payments by international tourists during their stay in the country (dela Cruz, 2023). Taking an active role, the DOT supports LGUs in digitalization-related projects for tourism for their respective localities (DOT, 2021) and launched a "challenge grant" for LGUs to craft tourism programs and projects that may benefit the tourism sector at large (Cruz, 2023). However, despite increasing optimism from the consumer side of the economy, the business sector at large remains relatively skeptical in economic prospects in the short run to the medium term. This is reflected in the indexes of the BSP for consumer expectation surveys (CES) and business expectation surveys (BES) conducted (Fig. 5.2, left panel).

Additionally, there must be emphasis that not only the private sector in supporting the totality of tourism industry but also the national government who ensures that such sector contributes to the macroeconomy and a source of value-added and employment. Note that the share of tourism sector to total investments and government spending are on a decline, which is in contrast with

[1]Other food service providers within hotels and resorts are covered in this item.

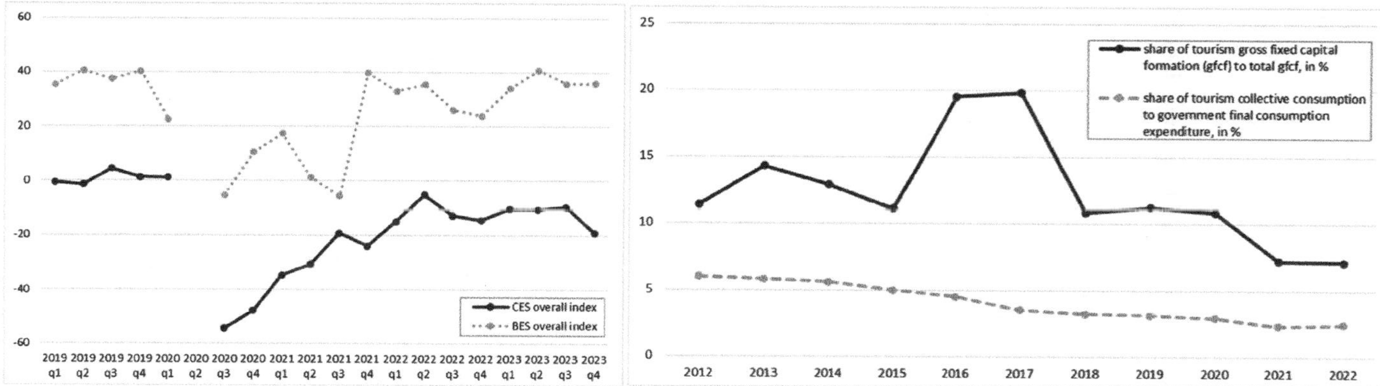

Fig. 5.2. [Left] BES and CES Overall Indexes, First Quarter of 2019 to Third Quarter of 2023.[2] [Right] Share of Investments in Tourism and Government Spending in Tourism to Total Investment and Government Spending, Respectively, Both in Percent, 2012–2022. *Source:* BSP and DOT.

the trend of domestic and inbound tourism (Fig. 5.2, right panel). It must be emphasized at this point that the collective contribution of various stakeholders in the tourism sector – public and private, domestic and international – will have rippling effects not only to businesses and employment in the sector but will also benefit the macroeconomy and will help realize the intended role of the sector.

Afterword: The Pandemic as a "Silver Lining" to Improve Local and Micro-Tourism Initiatives

The pandemic has highlighted the viability of local and micro-tourism as a resilient alternative approach that played a crucial role in sustaining the tourism industry during the crisis. Central to this approach is the active participation of MSMEs. To maximize the benefits derived from this new model, it is imperative to ensure that MSMEs in the sector survive and thrive. This is what several roadmaps from different international organizations such as Asian Development Bank (ADB) and the Southeast Asia Development Solutions (SEADS) aim to do (Tan, 2022).

ADB (2021) sees market-based financing in the form of capital markets and digital platforms as a critical part of the long-term survival of MSMEs. While financial assistance from the public sector has helped in the survival of MSMEs, increased market-based financing will improve the ability of enterprises to meet their capital needs and transition to digital platforms. Tan (2022) echoes this sentiment, highlighting digitalization as a key factor contributing to MSMEs' growth. According to the SEADS platform, digitalization (1) drives efficiency by allowing businesses to be productive with fewer employees, (2) improves customer reach by providing online-to-offline platforms and creating online markets, and (3) facilitates access to finance by providing an alternative source of data that allows financial institutions assess MSMEs' credit risk (Tan, 2022).

However, we must recognize that MSMEs face different challenges that affect their decision to follow these roadmaps. Thus, policymakers need to take active measures to close this gap, ensuring comprehensive growth for MSMEs. Acopiado et al. (2022) investigated the factors affecting Philippine MSMEs' adoption of digital payments in their operations, which indicated that firms are more inclined to adopt digital payments are firms that are younger, have a partnership business structure, have more employees, have existing ICT-based facilities, and experienced business growth prior to the pandemic. This is consistent with the uneven appreciation, access, and adoption to technology given the accelerated shift to digitization induced by the COVID-19 pandemic (Abueg, 2020b). Such need for digitization is a theme and highlight of the postpandemic roadmaps, especially for tourism sector resilience, recovery, and renewed sustainability (ADB, 2023), not only for pandemic-induced economic shocks but also on disaster-related vulnerabilities (Cambridge Industrial Innovation Policy, 2022).

---

[2]Note that no surveys were conducted during the second quarter of 2020, at the height of the lockdowns (BSP, 2023a, 2023b).

As the economy reopened, the growing confidence among consumers and businesses has had a positive trend since the pandemic, as illustrated in Fig. 5.2 right panel. This confidence is expected to translate into increased economic activities, particularly in the tourism sector, further enhancing the overall confidence of consumers and businesses. However, lingering uncertainty persists, with a significant portion of businesses expressing low expectations for economic improvement and fluctuations in overall consumer expectations.

As economic theory and development policy and studies literature point out, the government must be the lead in bringing the economy – prioritizing the vulnerable and severely affected sectors – out of the shocks of the pandemic and related economic recession effects. In particular, the government should be steadfast in supporting MSMEs to boost confidence as they play a vital role in revitalizing the country's tourism industry. Financial support and business process facilitation are key identified roles that the government must extend to microenterprises to ensure recovery and resilience, which may be hinted in Fig. 5.2. Additionally, a significant number of the microenterprises that are informal in nature would require intensified support to adopt to digitization and new methods of the next normal (ADB, 2021; Raquiza, 2021; UNCTAD, 2021). This support will augment the unintended inequality caused by technology adoption across sectors (Abueg, 2020b, 2023b). Such also recognizes the unequal and uneven recovery experiences of the economic sectors given the unequal access to the support of both the civic society and the public sector, resources for business, and technology adoption. Finally, as we ensure our tourism sector is adaptive to the changes of the "next normal," it must be given full recognition of the economic scarring that the pandemic has done to a lot of sectors. These adaption efforts must be taken into context with the persistent economic uncertainties and also the geopolitical events that have in a way influenced the economic trajectory of development (Debuque-Gonzales et al., 2022; Jocson, 2023): both for the whole economy, the tourism sector, and the microenterprises. With digitization becoming the thread of new models postpandemic, given the potentials of the microenterprises, there must be adequate support to these business entities to public programs and private initiatives, particularly in finance, technological support, and human investments (ADB, 2020b). In this way, with the small-scale movements, models, and innovations, we realize maximal gains and realize greater potentials. These are also aligned and supportive of a number of Sustainable Development Goals (SDGs) of the United Nations. Hence, as tourism gets small due to the pandemic experiences and lessons, we go big in innovation, developments, and opportunities.

References

Abueg, L. C. (2005). The economics of secondhand retail trade: An analysis of the market for *ukay-ukay*. *Philippine Journal of Development*, *32*(1), 53–77.

Abueg, L. C. (2020a). *Extended, enhanced, and extreme: Macroeconomic implications of the community quarantine in the Philippines due to the COVID-19 pandemic*. Economics and Management Matters no. 2020-01. University of the Philippines Los Baños.

Abueg, L. C. (2020b). Silver linings in Philippine history and macroeconomics of the Covid-19 pandemic response: Beyond the longest lockdown. *Philippine Journal of Health Research and Development*, *24*(4), 50–61.

Abueg, L. C. (2021). What can gender economics learn from the Pinoy BL genre of the Covid-19 pandemic? *Review of Women's Studies*, *31*(1), 35–62.

Abueg, L. C. (2023a). Stretching Filipino "resilience" amidst a pandemic-induced recession with market and government failures: Insights from community pantries and online barter. *Graduate Inequality Review*, *2*, 68–76.

Abueg, L. C. (2023b). ASEAN integration and tourism: Towards a sustainable and inclusive regional tourism program. In J. P. R. JPR Rivera, M. C. L. S. MCLS Rodolfo, F. M. Y. Roxas, & E. L. M. Gutierrez (Eds.), *People, planet, profit: Principles and practices of sustainable tourism* (pp. 431–464). Joint publication of the Asian Institute of Management (Makati City), & De La Salle University Publishing House (City of Manila).

Abueg, L. C., Largoza, G. L., Sauler, M. M. R., & Teehankee, B. L. (2016). The impact of a common good model of the firm on prosocial business decision-making of students: An experimental investigation. *Proceedings of the DLSU Research Congress, 4*.

Abueg, L. C., Sauler, M. M. R., & Teehankee, B. L. (2014). Towards a common good model of the firm. *DLSU Business and Economics Review*, *24*(1), 1–12.

Acejo, I. L., & Abueg, L. C. (2022). Sailing through post-pandemic recovery? The role of the state for workers in cruise tourism industry in the Philippines. In P. Mohanty, A. Sharma, J. Kennell, & A. Hassan (Eds.), *The Emerald handbook of destination recovery in tourism and hospitality* (pp. 227–248). Emerald Publishing Limited.

Acopiado, I. M., Sarmiento, J. M. P., Romo, G. D. A., Acuña, T. R., Traje, A. M., & Wahing, G. D. (2022). Digital payment adoption during the COVID-19 pandemic in the Philippines. *Philippine Journal of Science*, *151*(3), 1185–1196.

ADB. (2020a). *Asia small and medium-sized enterprise monitor 2020*. https://doi.org/10.22617/TCS200290-2

ADB. (2020b). *Asia's journey to prosperity: Policy, market, and technology over 50 years*. https://doi.org/10.22617/TCS190290

ADB. (2021). *More growth capital needed for micro, small, and medium-sized businesses in South Asia post-pandemic*. https://doi.org/10.22617/SGP210459-2

ADB. (2023). *Asia small and medium-sized enterprise monitor 2023: How small firms can contribute to resilient growth in the Pacific post COVID-19 pandemic*. https//doi.org/10.22617/SGP230413-2

Adel, R. (2021). DOT chief welcomes community pantries at tourism sites. *Philippine Star*. https://bit.ly/3Su0wbb

Arnaldo, M. S. F. (2020, August 20). Data speaks: Nearly all tourism outfits are small, medium, micro enterprises. BusinessMirror. https://bit.ly/3St4Qak

Bowles, S. (2016). *The moral economy: Why good incentives are no substitute for good citizens*. Yale University Press. https://yalebooks.yale.edu/book/9780300230512/the-moral-economy/

BSP. (2023a). *Business expectations survey*. https://www.bsp.gov.ph/SitePages/MediaAndResearch/ConsumerExpectationSurvey.aspx

BSP. (2023b). *Consumer expectations survey*. https://www.bsp.gov.ph/SitePages/MediaAndResearch/ConsumerExpectationSurvey.aspx

Buonincontri, P., & Micera, R. (2016). The experience co-creation in smart tourism destinations: A multiple case analysis of European destinations. *Information Technology & Tourism, 16*, 285–315. https://doi.org/10.1007/s40558-016-0060-5

Cambridge Industrial Innovation Policy. (2022). *Beyond COVID-19: What can be done to enhance the resilience of MSMEs?* https://bit.ly/3S63Zen

Canadian Trade and Investment Facility for Development. (2021). *Final study report for the post-Covid-19 recovery plan for Asian tourism*. https://bit.ly/4944VH2

Catoto, R. N. (2020, April 3). Siargao folk revive barter system. *Mindanews*. https://bit.ly/42f8xDZ

Caynila, K. A. M., Luna, K. T., & Milla, S. A. A. (2022). The Philippine tourism sector amid the pandemic: Developments and prospects. *BSP Economic Newsletter no. 22*. https://bit.ly/3tQXxA8

Correa, L. N. V., & Abueg, L. C. (2020). The Philippines' *Balik Probinsya, Bagong Pag-asa* initiative and the real estate development sector during the COVID-19 pandemic: Some policy morals. *International Journal of Reconfigurable and Embedded Systems, 14*(S2), 156–176.

Cruz, M. (2023). DoT opens tourism champions tilt. *Manila Times*. https://bit.ly/493Bnt0

Dacanay, S. J. O., III., Hapitan, R. B., & Abueg, L. C. (2018). Macroeconomics and industry analysis. In R. R. Santos (Ed.), *Essentials of investments in the Philippine capital market* (3rd ed., pp. 305–359). Southville Global Education Network.

Debuque-Gonzales, M., Corpus, J. P. P., & Miral, R. M. L. (2022). *Macroeconomic prospects of the Philippines in 2022–2023: Steering through global headwinds*. Philippine Institute for Development Studies Discussion Paper no. 2022-32. https://bit.ly/47E22vy

dela Cruz, E. (2023). *Philippines to offer value-added tax refund to foreign tourists by 2024*. Reuters. https://reut.rs/4b2w3YX

DILG. (2021). *Safety seal: For business establishments*. https://bit.ly/3u3PMqu

DOT. (2021). *DOT TPB continues to support LGUs with digital transformation projects*. https://bit.ly/3SqOcZ7

DOT. (2022). *PH's significant recovery in travel and tourism hot topic in Routes Asia 2022*. https://bit.ly/3vSwox1

DTI. (2019). *2019 Philippine MSME statistics in brief*. [Data set]. https://dtiwebfiles.s3-ap-southeast-1.amazonaws.com/BSMED/MSME+2019+Statistics/2019+Philippine+MSME+Statistics+in+Brief.pdf

DTI. (2020). *2020 Philippine MSME statistics in brief*. [Data set]. https://dtiwebfiles.s3.ap-southeast-1.amazonaws.com/BSMED/MSME+2020+Statistics/2020+Philippine+MSME+Statistics+in+Brief.pdf

DTI. (2022). *2022 Philippine MSME statistics in brief*. [Data set]. https://dtiwebfiles.s3.ap-southeast-1.amazonaws.com/MSME+Resources/2022+Philippine+MSME+Statistics+in+Brief.pdf

Dy-Zulueta, D. (2023, April 20). 'Revenge travel' is real: Group travel rises by 180%; list of top 'barkada' getaways. *The Philippine Star*. https://bit.ly/3Oe79vE

Farsari, I. (2021). Exploring the nexus between sustainable tourism governance, resilience and complexity research. *Tourism Recreation Research, 48*(3), 352–367. https://doi.org/10.1080/02508281.2021.1922828

Ganti, L. (2020, October 16). Covid-19 is accelerating the adoption of disruptive technologies and creating business opportunity. *Forbes*. https://bit.ly/47Fxjhz

Gutierrez, E. L. M. (2021). Redefining sustainable tourism in COVID-19: A political-economy approach. *DLSU Business and Economics Review, 31*(2), 1–13.

ILO. (2012). *Toolkit on poverty reduction through tourism training package teaching notes module 3: Promotion and marketing in tourism.* ILO. https://www.ilo.org/wcmsp5/groups/public/—ed_dialogue/—sector/documents/instructionalmaterial/wcms_218329.pdf

ILO. (2021). *COVID-19 and employment in the tourism sector in the Asia–Pacific region.* ILO. https://www.ilo.org/asia/publications/issue-briefs/WCMS_827495/lang–en/index.htm

Japan Times. (2019, June 12). Sustainable tourism will ensure the Philippines is #MoreFunForever. *The Japan Times.* https://bit.ly/3SsYd8j

Jocson, L. M. J. C. (2023, February 9). New growth drivers seen needed after pandemic economic scarring. *BusinessWorld.* https://bit.ly/3tYVXMx

Laorden, N. L., Sarmiento, J. M. P., Romo, G. D. A., Acuña, T. R., & Acopiado, I. M. A. (2021). Impact of supply chain disruptions during the COVID-19 pandemic to micro, small and medium enterprises in Davao Region, Philippines. *Journal of Asia Business Studies, 12*(3), 568–586. https://doi.org/10.1108/JABS-05-2021-0216

Lee, P., Hunter, W., & Chung, N. (2020). Smart tourism city: Developments and transformations. *Sustainability, 12*(10), 3958. https://doi.org/10.3390/su12103958

Marasigan, L. (2019). 17 commercial airports in 3 years, 28 ongoing. *Business Mirror.* https://bit.ly/3Odhgkb

Marcus, L. (2022, May 15). Revenge travel: How vacation vengeance became a thing. *CNN.* https://bit.ly/4b3hWm5

NEDA. (2023). *Philippine development plan 2023–2028.* https://pdp.neda.gov.ph/philippine-development-plan-2023-2028/

Ocampo, R. (2021, May 6). Displaced Philippine travel workers get aid from community pantry. *TTG Asia.* https://bit.ly/3vKDYKi

OECD. (2020). *Rebuilding tourism for the future: COVID-19 policy responses and recovery.* https://www.oecd.org/coronavirus/policy-responses/rebuilding-tourism-for-the-future-covid-19-policy-responses-and-recovery-bced9859/

Official Gazette of the Philippines. (2018). *Republic Act no. 11165.* https://www.officialgazette.gov.ph/2018/12/20/republic-act-no-11165/

Padronia, E. K. H. (2023, May 8). Incentives await MSMEs for tourism growth. *SunStar.* https://bit.ly/3vM9url

Panzer-Krause, S. (2022). Rural tourism in and after the COVID-19 era: "Revenge travel" or chance for a degrowth-oriented restart? Cases from Ireland and Germany. *Tourism & Hospitality, 3*(2), 399–415. https://doi.org/10.3390/tourhosp3020026

Pecsek, B. (2018). Working on holiday: The theory and practice of workcation. *Balkans Journal of Emerging Trends in Social Sciences, 1*(1), 1–13. https://doi.org/10.31410/Balkans.JETSS.2018.1.1.1-13

PSA. (2020). *Labor force participation, employment, unemployment and underemployment rates by region October 2019.* [Data set]. https://psa.gov.ph/sites/default/files/iesd/2022-12/TABLE%25201%2520Labor%2520Force%2520Participation%252C%2520Employment%252C%2520Unemployment%2520and%2520Underemployment%2520Rates%2520by%2520Region%2520October%25202019_0.xls

PSA. (2021). *Labor force participation, employment, unemployment and underemployment rates by region October 2020*. [Data set]. https://psa.gov.ph/sites/default/files/iesd/2022-12/TABLE%25201%2520Labor%2520Force%2520Participation%252C%2520Employment%252C%2520Unemployment%2520and%2520Underemployment%2520Rates%2520by%2520Region%2520October%25202020_0.xls

PSA. (2022). *More than 1.08 million establishments operated in 2021 which generated total employment of 8.57 million (2021 updating of the list of establishments preliminary results) (2022-07)*. [Data set]. https://psa.gov.ph/content/more-108-million-establishments-operated-2021-which-generated-total-employment-857-million

PSA. (2023). *2020 survey of tourism establishments in the Philippines (STEP) – Economy-wide: Preliminary results (23SSO003)*. [Data set]. https://psa.gov.ph/content/2020-survey-tourism-establishments-philippines-step-economy-wide-preliminary-results

Raquiza, M. V. R. (2021). *Micro, small, and medium enterprise (MSME) sector financing issues and challenges*. University of the Philippines Center for Integrative and Development Studies Discussion Paper no. 2021-01. https://cids.up.edu.ph/wp-content/uploads/2022/02/UP-CIDS-Discussion-Paper-2021-01.pdf

Rico, J., & de Leon, K. R. C. (2017). Mall culture and consumerism in the Philippines. *State of Power 2017*. https://www.tni.org/files/publication-downloads/stateofpower2017-mall-culture.pdf

Rocamora, J. A. L. (2021, December 28). *2021: Renewed hope towards 'steady' tourism recovery*. Philippine News Agency. https://www.pna.gov.ph/articles/1163894

Rocamora, J. A. L. (2022, February 10). *Support establishments with safety seal 'badges': DOT*. Philippine News Agency. https://www.pna.gov.ph/articles/1167470

Shinozaki, S., & Rao, L. (2021). *COVID-19 impact on micro, small, and medium sized enterprises under the lockdown: Evidence from a rapid survey in the Philippines*. ADBI Working Paper Series no. 1216. https://www.adb.org/publications/covid-19-impact-msme-under-lockdown-evidence-rapid-survey-philippines

Tan, M. (2022). Realizing the potential of over 71 million MSMEs in Southeast Asia. *ADB Southeast Asia Development Solutions*. https://seads.adb.org/solutions/realizing-potential-over-71-million-msmes-southeast-asia

Tayag, C. (2022, April 28). DOT reopens Philippine tourism with a big bang. *The Philippine Star*. https://bit.ly/426gHyy

TESDA. (2017). *The Philippine maritime industry through the year a labor market intelligence report*. https://www.tesda.gov.ph/Uploads/File/planning2017/LMIR/MARITIME%20LMIR%20March%2017%20with%20cover%202017.pdf

UNCTAD. (2021). *The COVID-19 pandemic impact on micro, small and medium sized enterprises*. https://unctad.org/system/files/official-document/ditcclp2021d3_en.pdf

Wang, J., & Xia, L. (2021). Revenge travel: Nostalgia and desire for leisure travel post COVID-19. *Journal of Travel & Tourism Marketing, 38*(9), 935–955. https://doi.org/10.1080/10548408.2021.2006858

Wiedmann, T., Lenzen, M., Keyßer, L. M., & Steinberger, J. K. (2020). Scientists' warning on affluence. *Nature Communications, 11*, 3107. https://doi.org/10.1038/s41467-020-16941-y

World Bank Group (2019). *Philippines: Assessing the effectiveness of MSME and entrepreneurship support*. https://documents1.worldbank.org/curated/en/

853041563828559514/pdf/Philippines-Assessing-the-Effectiveness-of-MSME-and-Entrepreneurship-Support.pdf

World Bank Group. (2022). *The global findex database.* https://www.worldbank.org/en/publication/globalfindex/Data

Zaman, U., Raza, S. H., Abbasi, S., Aktan, M., & Farías, P. (2021). Sustainable or a butterfly effect in global tourism? Nexus of pandemic fatigue, COVID-19-branded destination safety, travel stimulus incentives, and post-pandemic revenge travel. *Sustainability*, *13*, 12834. https://doi.org/10.3390/su132212834

Zhao, K. (2022). Research progress and prospect of micro-tourism scenarios. *Frontiers in Business, Economics & Management*, *4*(3), 15–18. https://doi.org/10.54097/fbem.v4i3.1045

Chapter 6

"Thinking Small": Exploring the Opportunities of Tourism During the Post-COVID-19 Era Using the Systems Thinking Paradigm

Daryl Ace V. Cornell[a], *Ethelbert P. Dapiton*[b] *and Liwliwa B. Lagman*[c]

[a]Polytechnic University of the Philippines, Philippines
[b]Don Honorio Ventura State University, Philippines
[c]University of Mindanao, Philippines

Abstract

Emerging from the COVID-19 pandemic, the Philippines has undergone the "new normal" transition, creating a strategic recovery effort to reinvigorate the industry. In tourism, these transitions aim to safeguard employees' and guests' health and safety, ensure continuity of business operations, boost tourism confidence leading to satisfaction, and establish a resilient and sustainable tourism industry in the postpandemic era. Hence, this chapter employs a system thinking leveraging a causal loop diagram (CLD) to construct a comprehensive roadmap for Philippine tourism's postpandemic resurgence through the system thinking lens. The CLD visually illustrates the inter-related factors influencing the recovery process, encompassing collaborative engagements, innovations, economic revitalization, and health and safety protocols. By analyzing the causal relationships among these variables, this chapter explicates the dynamic and interconnected nature of the postpandemic recovery leading to the recovery of the Philippine tourism industry, especially in the context of thinking small. Through this chapter, thinking small could involve a shift toward localized solutions and community-focused initiatives that allow them to foster local economies, build resilience, and create a more inclusive and sustainable postpandemic recovery.

Revisiting Sustainable Tourism in the Philippines, 107–134

Published under exclusive licence by Emerald Publishing Limited
doi:10.1108/978-1-83753-678-820241006

Keywords: Tourism; COVID-19; postpandemic; causal loop diagram; systems thinking; roadmap; plan and recovery

Introduction

Tourism is one of the backbones of modern economies that has an extreme effect on various aspects of a country's economic performance, not only boosting incomes and employment but also improving living standards (Agaraj & Murati, 2009). Tourism receipts demonstrate its positive economic effect, with the industry contributing 12.7% to the Philippine gross domestic product (GDP) and employing 5.7 million Filipinos (Canton, 2021a). Moreover, tourism aligns with the country's Sustainable Development Goals, offering potential for sustainable development, particularly evident during the COVID-19 pandemic.

However, the pandemic abruptly halted years of tourism industry progress. In 2020, global international tourist arrivals plummeted by 74%, impacting regions worldwide, including the Philippines, with estimated long-term costs totaling 41.4 trillion pesos and 4.3 trillion pesos lost during the peak of the pandemic (Canton, 2021b). This crisis exposed the industry's vulnerability, especially for micro-, small-, and medium-sized enterprises (MSMEs), leading to unemployment, deteriorating working conditions, and business closures (Canton, 2021a). Government interventions, including strict quarantine and health protocols, triggered the decline in travel and tourism-related activities (World Bank, 2022).

To address these challenges, the Department of Tourism (DOT) formulated the Tourism Response and Recovery Plan (TRRP) in April 2021, aiming to protect workers, sustain business continuity, and facilitate a resilient and sustainable industry revival postpandemic (DOT, 2021). This comprehensive strategy involves immediate response and relief efforts for affected tourism workers, as well as upskilling and modernization initiatives to adapt to new operational norms and health protocols, particularly in domestic tourism revival. Similarly, we would address the specific objectives:

- To utilize systems thinking (ST) in exploring linkages between health and safety protocols to the guests' satisfaction in response to the post-COVID-19 pandemic.
- To develop a causal loop diagram (CLD) that shows how the tourism industry responds to the health and safety protocols implemented that result in guest satisfaction.
- To identify intercession that leads to the hastening of the tourism industry's recovery and guest satisfaction during the post-COVID-19 pandemic.

By employing an ST approach, this chapter will analyze the connection of variables of government interventions, health and safety protocols, collaborations, innovations, economic revitalization, and resilience (Schiuma et al., 2012).

A CLD will visually represent the cause-and-effect relationships among these variables, aiding in understanding dynamic linkages and patterns within the system (Schiuma et al., 2012). Through this chapter, the tourism industry can identify effective interventions to facilitate industry recovery and development in the post-COVID-19 setup, guiding operational businesses, employees, and customers toward recovery and long-term success by enabling adaptation and innovation, ensuring sustainable growth.

Literature Review

Tourism Industry Recovery Through the Lens of ST

Applying ST in the Tourism Industry

ST, which originated from Bertalanffy (1968) systems theory, is a way of analyzing every corner of system relationships, perspectives, and boundaries critically to aid how we can deal with changes and make decisions (Baggio, 2008, as cited by Glyptou, 2022). ST assumes that understanding a system requires an analysis of its constituents and their interlinkages and relationships (Schiuma et al., 2012). Arnold and Wade (2015) described ST as the use of analytic skills to identify and understand a system, predict behaviors, and provide changes and improvements to be able to produce a better and desired outcome within a system. Consequently, ST involves broadening the perspective to encompass multiple linkages instead of keeping components within a studied system constant (Roxas et al., 2021).

The use of this method has been increasing in studies in different fields, such as sustainability science (Bai et al., 2016; Gretter et al., 2018; Williams et al., 2017), education (Tamim & Bostwick, 2020), and business management (Inghels, 2020). These industries are as complex as the area of tourism. Therefore, ST provides a framework for a deeper understanding of the issue, enabling the development of policy interventions designed for dynamically complex systems like tourism (Aronson, 1996, as cited in Roxas et al., 2018).

Several studies proved that ST is reliable for conducting studies, specifically in the tourism industry. For instance, Jaaron et al. (2021) conducted a study on tourism companies where ST was applied to knowledge-based structure realization, which is a survival planning tool in times of crisis such as the pandemic and could contribute to the growth of the economy and sustainable practices of tourism enterprise. It has also been used to investigate the underlying system structures that can influence sustainability, where the connection between the ST approach and tourism development has been identified (Mai & Smith, 2015). Similarly, in wildlife tourism, the method has been used as a lens to study local community development outcomes (Stone & Nyaupane, 2018). It has also been used to create an evaluation index system for urban tourism sustainable development (Lei et al., 2014). The approach has helped simplify and clarify specific problems within the tourism industry, which proves its significance in decision-making and learning (Van Mai & Maani, 2009).

Thinking Small in the Post-COVID-19 Pandemic

The COVID-19 pandemic significantly affected tourism, making it one of the earliest and hardest-hit industries. The Asia–Pacific region was among the first to experience the effects of COVID-19, resulting in a staggering 98% decline in tourist arrivals by October 2020. This crisis revealed the industry's vulnerability to global crises. When news of COVID-19 spread, travelers began promptly canceling or postponing their trips. Postpandemic situations like this indicate that it might take longer for businesses and tourism activities to resume their operations in the tourism scenario before the pandemic (Aburumman, 2020).

The pandemic has profoundly influenced the tourism sector and its associated industries, leading to a downturn in the Philippines' overall economic performance, household incomes, and livelihoods. It is paramount to delve into the implications of these repercussions, as this analysis serves as a key for a comprehensive evaluation of the efficacy of the measures and strategies executed during the pandemic and in its aftermath. Confident researchers suggested conducting further studies, such as looking into systemic approaches in addressing crises and disasters (Ritchie & Jiang, 2019; Sigala, 2020) and exploring complex topics by going beyond descriptive research, as research on tourism and COVID-19 should not solely focus on the aspects of the pandemic but instead building an interconnected perspective of the recovery plan (Zenker & Kock, 2020).

Given that the tourism industry has a reputation for resilience in the face of various environmental, socioeconomic, and political challenges, the challenges posed by COVID-19 suggest the need for profound structural changes in the industry (Sigala, 2020, as cited in Roxas et al., 2021). According to Gutierrez et al. (2020), it is necessary to adjust and adapt from the pandemic lessons, specifically by starting and serving the tourist in a sustainable and different way called "thinking small." Sectors within the tourism enterprise, such as accommodations, transportation, alimentation, security, and sanitation, may be strategically redistributed based on their variety. The pandemic has forced businesses to work with a broken workforce, which hinders and limits the productivity of the enterprises (Kumar et al., 2020). Despite this, the consumers focused on the impacts of their actions on the environment, adopted green movements, and sought more authentic and immersive experience that led to the growth of their will to have a more sustainable travel behavior, like traveling within their home's vicinity and avoiding populated spots (Seabra & Bhatt, 2022).

Tourism is a complex system (Mai & Smith, 2015) described by its components and stakeholders' relationships that are nonlinear (Baggio, 2008). In connection with this, natural calamities, evolving tourist patterns, and advancements in technology are the external factors influential in the development path of the tourism industry (Roxas et al., 2021). Hence, ST is a powerful tool that could explore the tourism industry's complexities, economy, environment, and local communities (Walker et al., 1998) and understand the opportunities of the tourism industry while operating in small operations during the post-COVID-19 pandemic.

Given the global and national impact of the pandemic, the tourism industry's resilience depends on the actions and decisions taken by the government to address public health concerns. The policies and legislation enacted will directly shape tourists' behavior regarding their health, safety, and security. The industry's recovery will manifest through various adjustments and transformations people make in response to the crisis. For example, new practices like social distancing and frequent hand sanitizing have become integral to tourists' lives (Baraero-Era & Del Rosario, 2020). Consequently, despite an ongoing health crisis, tourism is bound to recover as the government and the people work together to implement effective strategies.

In connection with this, ST allows the dynamic adaptation of a destination's growth limits when in crisis (Glyptou, 2022). It provides a model for change management through understanding these complex systems. It helps decision-making by being able to foresee the long-term impacts of every action, including the unintended costs of policies and strategies (Nguyen & Bosch, 2013).

ST will help the government, tourism enterprises, and policymakers grasp the dynamic nature of tourism (Roxas et al., 2018). As it allows the integration of various principles into other disciplines through ST, there will be more comprehensive understanding of the COVID-19 pandemic's impact on the industry and will allow policymakers to develop interventions through the recovery process of tourism (Roxas et al., 2021). This approach can be utilized to identify critical drivers in the tourism industry that can produce intended solutions during a crisis like the pandemic (Roxas et al., 2018).

Causal Loop Diagram

CLDs are illustrations that describe cause–effect relationships, providing an insight of how different parts of a system connect and differentiate with each other (Schiuma et al., 2012). Researchers have used CLDs in studies related to tourism with topics such as identifying leverage points (Roxas et al., 2021), evaluating the complexity of policymaking (Crabolu et al., 2023), and planning of sustainable tourism models (Husain et al., 2020; Ropret, 2014). CLD is relevant to understanding the complex systems in the tourism industry given that it has intricacies within its sectors and districts. It allows us to understand how complex policymaking can be and helps us evaluate their effectiveness when implemented (Crabolu et al., 2023). Understanding the relationships of these variables is necessary to understand how the tourism industry can recover during the era of postpandemic which can help in designing interventions and recovery plans (Roxas et al., 2021).

The construction of CLDs starts with identifying the variables and their causal relationship to each other. To label these connections, direct proportional changes are marked with "+," while inverse relationships are labeled with "−." Lannon (2012) described CLDs by having two types of loops: Reinforcing (R) and Balancing (B). Reinforcing loops are labeled with an even number of "+/−," which indicate that a change in one direction within its arrows is amplified by a change. Meanwhile, Balancing loops are characterized by an odd number of "+/−" and indicate that a change in one direction in its arrows can be counteracted with a change in the opposite direction. However, it is important to note that systemic effects created in the CLDs will manifest only after undergoing multiple cycles (Roxas et al., 2021). The results comprehended from the diagram will serve as an instrument to help describe the complex cause–effect linkages and relationships among variables (Schiuma et al., 2012) in the tourism industry.

Hence, this chapter aims to contribute to the existing literature on the Philippine tourism industry by constructing a model that would present the complexity of the pandemic – an in-depth analysis of policies implemented through CLDs and possible interventions for unexpected consequences. In turn, it will contribute to sustainable efforts, reduction of economic leakage, and promotion of benefits and values equal distribution (Hall, 2009 as cited by Roxas et al., 2021) incorporating the new way of serving the tourists, "thinking small."

Research Gap

The assessed publications highlight the importance of ST in analyzing the impact of the COVID-19 pandemic on the Philippine Tourism Industry. This approach involves examining multiple levels, from mitigation policies to interventions involving various stakeholders and extending to broader systemic factors. To effectively revive and sustain the tourism industry postpandemic, a comprehensive understanding of this complex system is necessary. This involves utilizing CLDs to design a sustainable tourism framework, with a focus on empowering local communities and small enterprises to address sustainability challenges.

Methods

This chapter utilized a qualitative approach using a descriptive-historical research design to answer the research gap. We mainly hinge on analyzing primary data of the Philippine government's TRRP, by the DOT, dated June 2021. Using ST, we will develop CLDs designed to explain stakeholder revival responses and explore a path of revival and opportunities in the post-COVID-19 pandemic for a sustainable Philippine tourism industry to function amid various calamities.

Findings

CLD On-Process

The ST framework begins with a CLD of how sustainable tourism works. The baseline CLD denoted as a balancing loop (B1) represents the fundamental process underlying the unprecedented result of the COVID-19 pandemic on the Philippine Tourism Industry and the potential drivers via which tourism recovery, revival, and expansion can be facilitated. Fig. 6.1 shows that during the pandemic, government intervention, such as the Philippine government, plays a vital role in the tourism industry innovation (Mei et al., 2010). As such,

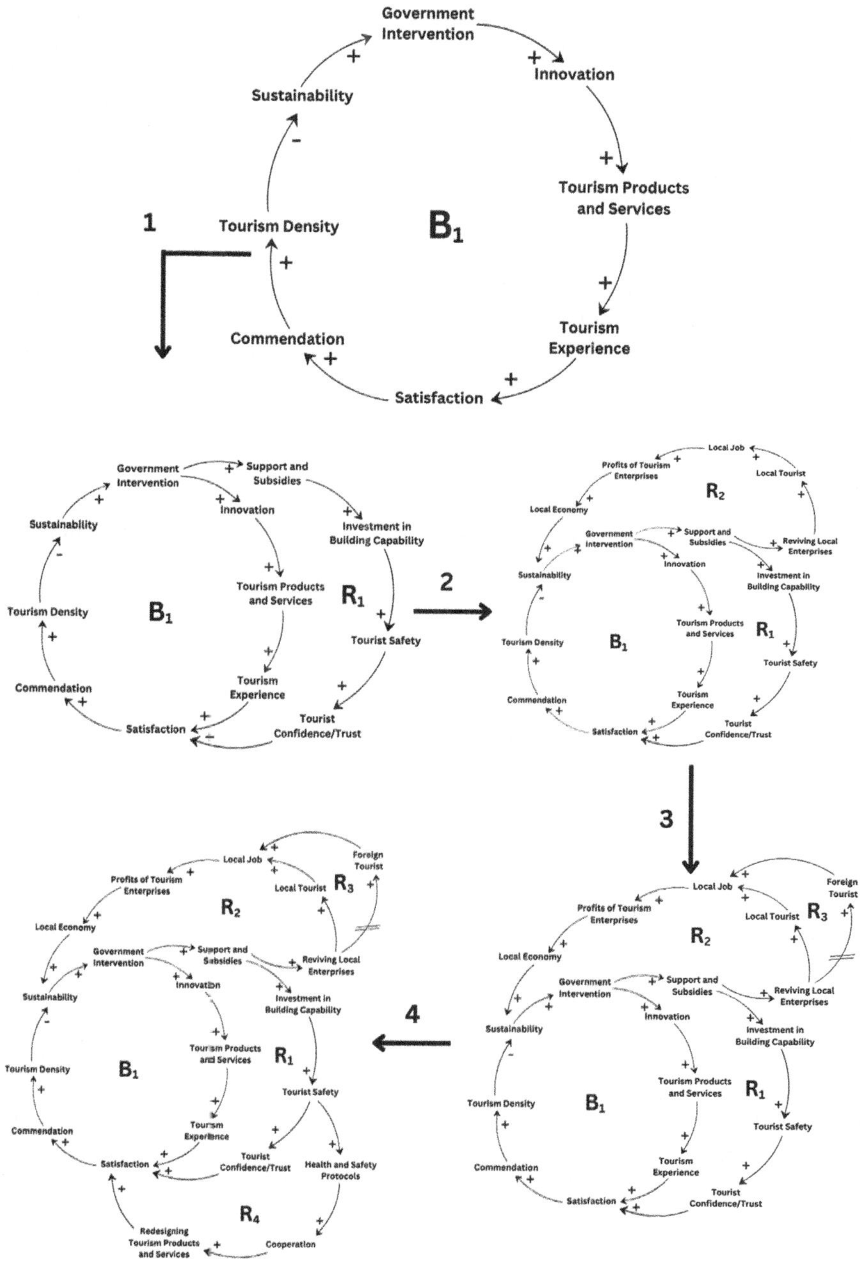

Fig. 6.1. CLDs – Author Constructed. *Note:* Systemic developments in B1, R1, R2, R3, R4 exemplify after multiple times. Immediate effects do not necessarily accrue.

innovations have been made and inextricably linked to tourism products and services advancements to mitigate (Li et al., 2023), which results in positive and favorable tourism experiences for tourists (Paasovaara et al., 2012; Tam, 2008). The enhanced tourism experience contributes to increased tourist satisfaction and adopting sustainable tourism practices (Huete-Alcocer et al., 2019; Torabi et al., 2022). Through formal tourism promotions, this will reap recognition from tourists quickly and from the government (Chang, 2014; Sigala, 2020). With this, the government's objective of attracting more tourists increases tourism density (Torabi et al., 2022). Maintaining an optimal balance between tourist influx and the resident population is crucial for the sustainability of the tourism industry (Upadhaya et al., 2022). This leads to the government's formulation of sustainable tourism policies, meeting contemporary society's demands while safeguarding future generations' interests (Amoako et al., 2021). From here on, communities are already secure enough to return to their everyday lives, maintaining the critical lesson from the pandemic that health and safety and sustainable tourism sustain.

Sustainable tourism is crucial in protecting natural resources, safeguarding cultural legacies, and promoting economic development within local communities (Upadhaya et al., 2022). The enhancement of resource utilization, community empowerment, and the provision of a gratifying visitor experience are all actively advocated for. Destinations can enhance their financial streams and standing and mitigate the impacts of climate change through adopting sustainable practices (Costa et al., 2021). Sustainable tourism is essential in ensuring the long-term sustainability and viability of the tourism business (Torabi et al., 2022). A practical and morally sound tourist enterprise is underpinned by a comprehensive plan that prioritizes the welfare of every party involved (Costa et al., 2021).

The CLD is expanded with a reinforcing loop (R1). R.A. 9593 (Tourism Act of 2009) empowers the government to establish a national tourism action plan, fostering collaboration among various sectors (Republic of the Philippines, 2009). It strengthens the capacity of local government units (LGUs) through partnerships with private sectors, facilitating effective management of tourism projects. Additionally, R.A. 9593 supports the creation of tourism enterprise zones, facilitating development in designated areas.

During the COVID-19 pandemic, the Philippine government provided support and subsidies, such as strengthening health response capacity and vaccination (World Bank, 2022). DOT invested in building tourism capability in regions and supported sustainable tourism development projects like Coron's Sustainable Tourism Development Project (STDP), accelerating the area's recovery (Department of Tourism, 2021). These efforts align with the trend of sustainable tourism, crucial during and after the pandemic (Azzahra, 2023). DOT also ensures support for sustainable tourism in known tourist spots (DOT, 2021).

Government interventions aimed at controlling the spread of COVID-19 and mitigating its consequences have been effective (Ashraf, 2020; Won et al., 2019). Subsidies and policies have led to interventions in the tourism industry, including quarantine policies, economic assistance, and healthcare spending (International Monetary Fund, 2020). Public sector support takes various forms, such as grants and subsidized loans, aiding communities in planning and policy procedures, fostering collaboration, and addressing community concerns (Javier & Elazigue, 2011).

Furthermore, continued government support for tourism recovery includes budget increases for development projects. In 2023, DOT Secretary Christina Garcia Frasco requested a budget of P15.3 billion, including P17.689 billion for the Tourism Road Infrastructure Program (TRIP) to improve access roads to tourism destinations (Department of Budget and Management, 2023). The 2024 budget sees a further increase, supporting old and new projects for tourism sector recovery (Cao et al., 2023). Investing in capability building and resilience is crucial for the tourism industry's survival and growth amid challenges like the pandemic (Jiang et al., 2020).

Based on the Updated TRRP of the DOT (2021), government efforts have significantly improved service quality and safety in the tourism industry, attracting both domestic and international travelers. Ensuring excellent service and safety measures is crucial for attracting tourists (Kuznetsova, 2023). Hence, the Philippine government actively enhances service standards and safety practices, including reopening safe tourist destinations nationwide amid the pandemic (DOT, 2021).

Prioritizing tourist safety ensures a smoother and more enjoyable experience, increasing tourists' confidence to revisit the Philippines (Jong, 2022). This assurance fosters trust in service providers and encourages word-of-mouth endorsements and positive referrals, attracting more visitors (Dirgantara et al., 2023; Harris, 2014). Once tourist experience and satisfaction increase, it leads to generating more recommendations that captivate more local visitors to join in various tourism movements (Gaki et al., 2016). Compliance with the United Nations World Tourism Organization (UNWTO) and World Travel & Tourism Council (WTTC) guidelines enhances safety, security, and hygiene, creating a safe travel environment (Roxas et al., 2021). Focusing on visitor security and establishing confidence and trust systematically benefits the tourism sector.

The reinforcing loop (R2) expanded the CLD. Government intervention plays a significant role during the pandemic (Ng et al., 2011) and in the tourism sector's recovery through financial help and subsidies (Roy, 2021). Hence, support and subsidies exist because they are forms of government intervention (Van Heuvelen, 2023) and significant instruments of government expenditure that can affect resource allocation, income distribution, and expenditure productivity (Clements et al., 1995). For instance, one of the government initiatives amid the COVID-19 crisis is the Tourism Rehabilitation and Vitalization of Enterprises and Livelihood (TRAVEL) program. This collaborative effort involved the DOT partnering with small businesses, where both organizations signed a Memorandum of Agreement, creating Coronavirus Aid, Relief, and Economic Security (CARES) for the TRAVEL program (DOT, 2021).

The DOT published an updated TRRP in June 2021 highlighting the importance of tourism marketing (DOT, 2021). These subsidies from the local enterprises help them continue their operations which help them develop and improve their ability to survive in demanding situations (Zhou et al., 2023).

The collaboration between the government, enterprises, and the community helped revitalize the tourism industry and support local enterprises (Baloch et al., 2022).

As local tourism enterprises gradually bounce back from the repercussions of the pandemic, the number of local tourists consequently increases (Sharma et al., 2021). The feeling of belongingness in the local destination can encourage domestic tourism as individuals seek to explore their surroundings and heritage, which contributes to the revival of enterprises and an increase in local tourists (Baloch et al., 2022). For instance, a surge in domestic tourism in 2021 has been reported, recording 37,279,282 trips, marking a 38.16% increase compared to the previous year (DOT, 2021) after the tourism enterprise gradually returned to the industry.

The resurgence of local enterprises coincides with an increase in local tourists and has helped stimulate job creation (Roy, 2021). Creating local jobs is linked with the social and economic sustainability of a destination, including the amount and quality of jobs offered (Agyeiwaah et al., 2017). The surge in local employment within the tourism sector contributed 11.1% to the total employment rate of the Philippines (Statista, 2022). The profit from enterprises recirculates in the local economy, which benefits other businesses and the business itself (Yehia, 2019), proving that this economic profitability positively impacts and facilitates the local economy's recovery. A good movement in the local economy due to tourism-related enterprises will lead the local communities to participate in conservation (Bajracharya et al., 2006, as cited by Roxas et al., 2021) and promote social and cultural change (Sharma et al., 2021), which will consequently enhance sustainability (Roxas et al., 2018) and promote responsible tourism sustainable practices (Sharma et al., 2021).

The model is expanded with a third reinforcing loop (R3). Tourism is an engine for economic growth, associated with positive impacts such as stimulating domestic tourism, creating employment, and generating foreign exchange earnings (Steiner, 2006). Hence, the recovery of local enterprises is associated with a concurrent influx of foreign tourists. In 2022, a significant arrival influx of foreign tourists was recorded in the Philippines compared to the preceding year. Between February and December 2022, tourist arrivals increased by two million (Statista, 2022).

Additionally, community-driven tourism initiatives in local enterprises such as local craft markets and eateries can attract foreign tourists (Gul, 2023), which can contribute to revitalizing the tourism sector and economy, thereby increasing local jobs. The combined involvement of domestic and foreign tourists in local tourism will foster local economic growth and secure the local tourism enterprises' financial needs (Moaveni, 2014). Therefore, the increase in foreign tourists has led to the generation of local employment opportunities.

During the pandemic, tourist safety was one of the most important things being ensured by the public sector by developing health and safety protocols in the Philippines and globally. This has been done since health and safety protocols are deemed crucial during the pandemic to protect oneself and others from COVID-19 (Manulife Holdings Berhad, 2019). The health and safety protocols guidelines for reopening establishments amid COVID-19 are also publicly available and can be accessed by individuals. To further ensure tourist safety, The WTTC has developed global protocols and a stamp for the new normal to ensure that health, hygiene, and safety measures are consistently applied throughout the travel experience (WTTC, 2021).

To ensure the implementation of safety measures, it is essential to have community involvement and develop forward-thinking tourism guidelines that strike a balance between the quantity and quality of tourism (Gilmore, 2017). Similarly, effective collaboration of people in implementing and adhering to the health and safety protocols is vital to the process of prompt recovery of the industry (WTTC, 2021), which is why community participation has an important role in destination redesigning planning and implementation (Gilmore, 2017). The involvement of the local community in the process leads to the enhancement of the tourist experience, allowing them to engage in local customs and explore different travel perspectives (Bhandari, 2022). Therefore, community cooperation is integral for successfully redesigning tourism products and services (Krittayaruangroj et al., 2023). The ongoing impact of these redesign efforts promotes sustainable tourism (Roxas et al., 2021) and contributes to an overall increase in tourist satisfaction (Hu & Li, 2023).

Furthermore, this will also improve the destination's image, attracting more tourists (Sánchez-Sánchez et al., 2021), and will lead to a positive impact on tourist satisfaction (Zimik et al., 2022). This underscores the importance of considering tourist expectations and preferences in the redesign process to improve overall tourist satisfaction.

Full CLD

From the last set of CLDs in Fig. 6.1, it underscores the pivotal role of government intervention in the overall system. The balancing loop illustrates the government's potential to initiate successful innovations, thereby enhancing tourism products and services. Recognized factors influencing the tourism industry's success include delivering high-quality service and maintaining customer satisfaction (Stevens et al., 1995). This underscores the importance of regulating tourist density through innovative interventions in tourism products and services, leveraging technology to enhance the country's tourism portfolio (Agarwal, 2019; Roxas et al., 2018).

DOT's collaboration with other agencies, organizations, and ventures emphasizes effective tourism management and support. The revision of existing laws, such as the Tourism Act of 2009 and the TRRP (DOT, 2021), alongside new regulations, including the Sustainable Tourism Development Project, has hastened tourism recovery in those areas and continuously promoted sustainability in the destinations (Azzahra, 2023).

Another critical aspect in the Philippine tourism industry recovery is the restoration of tourist confidence through improvements in service standards and safety measures. To motivate tourists to travel, they must feel reassured about their safety at every stage of their journey (Jong, 2022). Enterprises and authorities must subsidize protective measures (Fotiadis et al., 2020) and develop secure travel plans, concrete health, and safety protocols (Gutierrez et al., 2020).

Considering consumer satisfaction and dissatisfaction hinge on postpurchase feelings, the country must present its tourist destinations as attractive places to attract domestic and international visitors (Hui et al., 2007; Solomon, 2002), utilizing available innovations.

In pandemic and postpandemic scenarios, traveler priorities emphasize safety, hygiene, and protection (Bonfanti et al., 2021; Hu et al., 2021). Another opportunity identified is how the tourism industry should continuously monitor and adapt in desperate situations. The intensified implementation of health and safety protocols during and after the pandemic reflects the acknowledgment of these priorities. Travelers are expected to prioritize health and safety in tourism activities, favoring natural areas, digital travel, customized experiences, and uncrowded destinations (Gutierrez et al., 2020). Hence, the systemic effects of prioritizing health and safety protocols and building trust and confidence among tourists will steadily increase as the proper implementation of effective industry-wide health and safety measures is necessary for the tourism sector's revival (Osewe, 2020). Moreover, adopting technological innovations during the pandemic to minimize physical contact between individuals (Kaushal & Srivastava, 2021) allowed the tourism industry to utilize technology toward a more sustainable approach while still achieving tourist satisfaction. As an answer to the severe consequence of the pandemic on countries like the Philippines, systemic improvement opportunities emerge (Kuznetsova, 2023). The Philippine government is actively enhancing service quality and tourist safety through supportive measures and subsidies, as reflected in the investments in building capability (R1) and the revival of local enterprise (R2).

The tourism industry revival (R1) is anticipated to lead to sustainability. Hence, it also shows how sustainable tourism can be a driver and an opportunity in the postpandemic era. Successful ecotourism enterprises create more local job opportunities, motivating stakeholders to adopt sustainable practices, thereby fostering ecotourism flourishment. It is crucial to recognize that improvements in tourism may lead to increased tourist arrivals, which could threaten sustainability promotion. Consequently, governmental responsibility lies in ensuring sustainability and mitigating unintended consequences that could undermine overarching sustainability goals (Roxas et al., 2018). Continuous efforts to promote sustainable tourism will positively affect tourist satisfaction (Hu & Li, 2023) and destination image improvement and consequently attract more tourists (Sánchez-Sánchez et al., 2021). However, it is noted that the influx of tourists could threaten sustainability (Roxas et al., 2018) as it could potentially exceed the maximum carrying capacity of a destination.

Given all the identified opportunities using the Sustainable Tourism paradigm, we presume that incorporating the "thinking small" idea is integral to reviving the tourism industry, eventually leading to sustainability. Despite imposed restrictions, the tourism industry's recovery highlights the resilience of small-scale operations. This suggests that tourists are willing to patronize establishments even if they are working in small operations, as long as their needs are met. Consequently, "thinking small" emerges as a potentially influential lever in promoting sustainability within the tourism sector.

However, it is important to emphasize that a transition to a more sustainable form of tourism necessitates collaborative efforts to restart the local economy through local tourism. Stakeholders, including tourists and governmental bodies, must collaborate to foster an industry that fosters "thinking small" (Gutierrez et al., 2020). This collaborative approach holds the potential to reshape the tourism landscape, aligning with sustainability goals and mitigating the issues posed by mass tourism and future challenges, such as the pandemic.

Empirical Evidence Supporting the CLDs

From Table 6.1, the paradigm was formed through the collection and analysis of empirical evidence from related literature. These studies support and justify the inter and causal relationship of the variables in the CLD, backing its reliability.

Conclusion

This chapter highlights the challenges faced by the tourism sector in the COVID-19 postpandemic, explicitly focusing on the Philippine tourism industry. It emphasizes the role of sustainability in addressing these challenges and introduces an ST approach to understand better and tackle the complexities involved. This chapter book visually presents the corresponding recovery phases and opportunities in postpandemic Philippine tourism through a CLD. The CLDs highlight the importance of government intervention, collaboration, safety protocols, domestic and international tourism, marketing strategies, economic growth, and sustainability as key factors influencing the industry's success. As these factors interact continuously, the Philippines aims to rebuild its status as a prominent tourist destination while preserving its natural and cultural assets.

The ST approach is positioned as an effective tool for addressing concerns about sustainable tourism. It allows for the identification of underlying systems influencing sustainable tourism development. We urge researchers to utilize this approach to identify critical factors and intervention areas, enhancing the effectiveness of proposed antidotes. ST also facilitated the integration of ideas from various specializations, enabling policies that proactively consider unintended effects and potential future issues. The application of ST can assist stakeholders in the tourism sector, including governance agencies, tourism planners, and other relevant parties, in understanding the industry's complex and dynamic nature. It provides a comprehensive perspective by streamlining and integrating seemingly disparate issues within the industry. Additionally, the methodology can be employed to assess and develop innovative tourist policies that consider the multifaceted challenges, interdependencies, and the inadequacy of simplistic solutions.

Table 6.1. Empirical Evidence.

CLD	Analysis	Source	Key Findings
B1	Government intervention plays a pivotal role in tourism's innovation.	Mei et al. (2010)	Government acknowledges the importance of innovation in tourism, but its efforts in collaboration, small business innovation, and funding are partially new. Government plays a passive role, with many innovation initiatives seen as the private sector's responsibility.
	Innovation has been inextricably linked to the advancement of tourism products and services.	Li et al. (2023)	The internet and information technologies play a vital role in tourism, including operations, structure, marketing, and strategy. E-tourism utilizes the internet to establish commercial connections, offering tourism-related services like flights, hotel bookings, and vehicle rentals, which enhances the overall efficiency and effectiveness of businesses in the industry.
	Tourism products and services result in a favorable tourism experience.	Tam (2008)	Favorable customer decision-making and promising behavior are linked to consumer acquaintance with brand by-products or services, such as first-time and repeat acquisitions.

Enhanced tourism experience contributes to increased tourist satisfaction.	Torabi et al. (2022)	Tourists' satisfaction and likelihood to revisit smart destinations are significantly impacted by their memorable experiences.
Tourist satisfaction reaps commendations.	Sigala (2020)	More heightened tourist satisfaction reaps recognition from visitors in the short run and from the administration through legal tourism publicity, which also garners formal support from the government in promoting tourism.
Commendations increase tourism density.	Torabi et al. (2022)	Policymakers and stakeholders can utilize commendations to formulate competent tourism plans that enhance visitors' experiences and the management of destinations.
Tourism density necessitates sustainability.	Upadhaya et al. (2022)	Maintaining an optimal balance between tourist influx and the resident population is essential for the sustainability of the tourism initiative.
Sustainability formulation demands government intervention.	Amoako et al. (2021)	Public stakeholders need to be more aware of formulating a sustainable tourism policy that intends to meet contemporary society's demands while safeguarding future generations' interests.

Continued

Table 6.1. *(Continued)*

CLD	Analysis	Source	Key Findings
R1	Government intervention implements support and subsidies.	Won et al. (2022), IMF (2020), and Ashraf (2020)	It is essential for public sectors to carefully design and execute subsidy strategies to maximize their positive effects and minimize their negative consequences. Several studies and reports have examined the effect of government interventions, including support and subsidies, during the COVID-19 pandemic. These interventions have encompassed many measures, such as quarantine policies, economic assistance packages, healthcare spending, and stringency measures. While the specific causal relationship between government intervention and its impact on the pandemic may vary by country and the nature of the interventions, the evidence suggests that public sectors' actions have a vital role in managing the increase of the virus and mitigating its consequences.
	Support and subsidies to investment in building capability.	Javier and Elazigue (2011)	The LGU's support and subsidies for investment in building capability can take various forms, such as grants, subsidized loans, and financial assistance. This can benefit the

			communities by providing ideas, authority, infrastructure, and policy and planning procedures, thus improving capability building.
	Investment in building capability for tourist safety.	Jiang et al. (2023)	Tourism organizations must invest in dynamic building capability and consider relying on the broader system to help them thrive and survive in environmental crises or disasters.
	Tourist safety to tourist confidence/trust.	Kuznetsova (2023)	When tourism personnel prioritize the tourists' safety and ensure they are well taken care of, they have a smoother and more pleasurable experience. This amplified focus on tourist safety has increased tourists' confidence to revisit the Philippines since they are more aware that its tourism personnel are well-trained and equipped with new learnings regarding fundamental health and safety protocols.
	Tourist confidence/trust to satisfaction.	Gaki et al. (2016)	When tourists have confidence and trust in their tourist service providers, it bolsters their participation and contentment, allowing an increase in commendations that further attract local tourists to join various tourism activities.
R2	Government intervention to support and subsidies.	Van Heuvelen (2023) and Schwartz et al. (1995)	The article discusses the growing global trade tensions caused by increased

Continued

Table 6.1. *(Continued)*

CLD	Analysis	Source	Key Findings
			subsidies from major economies, such as the US Inflation Reduction Act, the EU Green Deal Industrial Plan, and China's Made in China 2025 strategy. Problems arise concerning a possible backing war, emphasizing the importance of understanding the motivations behind these actions. Subsidies in terms of fiscal charges and damaging impacts on efficiency impose burdens on the economy.
	Support and subsidies to reviving local enterprises.	DOT (2021), Zhou et al. (2023), and Baloch et al. (2022)	The DOT and the Small Business Corporation have signed a Memorandum of Agreement for the CARES for Tourism Rehabilitation and Vitalization of Enterprises and Livelihood (TRAVEL) program to support MSMEs in the tourism industry for pandemic recovery. The study conclusions deliver an academic foundation for assisting policymaking during pandemics and trace the limitations of administration intervention in the community and the economy. Pakistan's tourism and hospitality sector has developed during COVID-19 and is

		poised to rejuvenate into an enduring enterprise.
Reviving local enterprises to local tourists.	Sharma et al. (2021)	Tourism industry might shift toward a renewed international economic demand marked by sustainable tourism, societal well-being, environmental action, and resident engagement.
Local tourist to local job.	Roy (2021)	The government must lead in revitalizing the tourism sector through financial stimulus packages and subsidies. Additionally, marketing activities, promotional tools, and social media platforms will contribute to the recovery process.
Local job to profits of tourism enterprises.	Agyeiwaah et al. (2017)	This study emerged with seven fundamental indicator themes of sustainable tourism: career creation, company viability, quality of life, water grade, waste management, energy saving, and community integrity supervision.
Profits of tourism enterprises to local economy.	Yehia (2019)	The article tackles tourism's substantial contributions to global economies through revenue generation, diverse employment opportunities, infrastructure development, and cultural

Continued

Table 6.1. *(Continued)*

CLD	Analysis	Source	Key Findings
			exchange between tourists and local communities.
	Local economy to sustainability.	Roxas et al. (2018) and Roxas et al. (2021)	A framework to highlight how ecotourism can create profitable businesses and local jobs and drive ecotourism to higher levels. Rebooting tourism is vital to rebooting the economy.
R3	Reviving local enterprises to foreign tourists.	Steiner (2006)	Understanding the growth of the Egyptian tourism sector requires considering its integration into the country's neo-patrimonial political system and its status as a semi-rentier state in the political economy.
	Foreign tourists to local job.	Gul (2023) and Moaveni (2014)	Local tourism catalyzes community empowerment and sustainable economic development, providing a beacon of hope amid global challenges. Overcoming obstacles contributes to building a resilient and inclusive tourism landscape. Macroeconomics factors did not significantly affect the profitability and financial performance of the tourism industry.

R4	Tourist safety developed and adopted health and safety protocols.	WTTC (2021)	The WTTC has invented multinational protocols and an emblem for the latest normal to ensure that health, hygiene, and security measures are consistently implemented throughout the journey.
	Health and safety protocols engage cooperation from travelers.	Gutierrez et al. (2020) and Gilmore (2017)	Following the critical outcomes from diverse travel surveys performed during the pandemic, tourists are anticipated to contend in tourism movements with health and security as their primary consideration, further suggesting their bias toward uncultivated spots and uncrowded destinations, digital trips, and customized ventures. Community participation is necessary to plan and implement initiatives in redesigning destinations.
	Cooperation is essential in redesigning tourism products and services.	Krittayaruangroj et al. (2023)	Cooperation is essential in redesigning tourism products and services.
	Redesigning tourism products and services leads to increased satisfaction.	Sánchez-Sánchez et al. (2021) and Zimik et al. (2022)	By redesigning tourism products and services, destinations can improve their image and attract more tourists, increasing satisfaction. It shows how vital it is for policymakers to be more mindful of the significance of tourism and infrastructure development.

Source: Author tabulated.

The exploration in navigating a path of opportunities for the post-COVID-19 era in Philippine tourism has advocated for an all-inclusive and systems-oriented approach to addressing the complexities of the postpandemic tourism landscape, specifically emphasizing sustainability in the Philippine tourism industry. It underscores the practical evidence indicating a positive trend in the tourism industry, suggesting a return to homeostasis. This positive course is attributed to reopening international borders and relaxing health and safety requirements. The ongoing immunization process is the root cause for these actions, leading to a decline in COVID-19 cases. Even with the easing of health and safety regulations, the Philippines has effectively managed to sustain sufficient capacity within its healthcare system and maintain low rates of hospital bed utilization. This implies that the healthcare infrastructure is resilient and capable of handling potential challenges even as pandemic-related restrictions are lifted. However, a warning is introduced, emphasizing that COVID-19 still poses a significant risk to specific subpopulations. This acknowledgment stresses the need to continue implementing public health efforts to safeguard vulnerable groups.

The industry's progression is attributed to key factors such as resilience, ingenuity, and a commitment to providing unforgettable experiences for visitors of all sorts. This positive outlook suggests that the tourism sector is adapting to the evolving circumstances, leveraging its strengths, and demonstrating a dedication to rebound and succeed in the postpandemic landscape. Considering the opportunities identified through the Sustainable Tourism (ST) approach, we believe that embracing "thinking small" is necessary to revitalize the tourism industry and achieve sustainability. This chapter emphasizes the need for adjustments in the tourism sector, learning from the pandemic, and introducing a new approach to serving tourists characterized by "thinking small." This involves strategically redistributing sectors like accommodations, transportation, food services, security, and sanitation based on their diversity. Despite restrictions, the observed recovery of the tourism industry underscores the resilience of small-scale operations. It indicates that tourists are willing to support smaller establishments if their needs are met. Therefore, "thinking small" emerges as a potentially powerful strategy for promoting sustainability in the tourism sector. We also suggest the utilization of public stakeholders on the proposed CLDs for when another pandemic happens to ensure minimal unfavorable effects on the industry.

References

Aburumman, A. A. (2020). COVID-19 impact and survival strategy in business tourism market: The example of the UAE MICE industry. *Humanities and Social Sciences Communications*, *7*(1), 1–11. https://doi.org/10.1057/s41599-020-00630-8

Agaraj, X., & Murati, M. (2009). Tourism an important sector of economy development. *Annals-Economy Series*, *1*, 83–90.

Agarwal, P. K. (2019). Retrieval practice & Bloom's taxonomy: Do students need fact knowledge before higher order learning? *Journal of Educational Psychology*, *111*(2), 189.

Agyeiwaah, E., McKercher, B., & Suntikul, W. (2017). Identifying core indicators of sustainable tourism: A path forward? *Tourism Management Perspectives, 24*, 26–33. https://doi.org/10.1016/j.tmp.2017.07.005

Amoako, G. K., Obuobisa-Darko, T., & Marfo, S. O. (2021). Stakeholder role in tourism sustainability: The case of Kwame Nkrumah Mausoleum and centre for art and culture in Ghana. *International Hospitality Review, 36*(1), 25–44. https://doi.org/10.1108/ihr-09-2020-0057

Arnold, R. D., & Wade, J. P. (2015). A definition of systems thinking: A systems approach. *Procedia Computer Science, 44*, 669–678. https://doi.org/10.1016/j.procs.2015.03.050

Ashraf, B. N. (2020). Economic impact of government interventions during the COVID-19 pandemic: International evidence from financial markets. *Journal of Behavioral and Experimental Finance, 27*, 100371. https://doi.org/10.1016/j.jbef.2020.100371

Azzahra, A. (2023). Implementation of good governance in public services at local government. *International Journal of Social Service and Research, 3*(7), 1899–1906.

Baggio, R. (2008). Symptoms of complexity in a tourism system. *Tourism Analysis, 13*(1), 1–20. https://doi.org/10.3727/108354208784548797

Bai, L., Huan, S., Gu, J., & McClements, D. J. (2016). Fabrication of oil-in-water nanoemulsions by dual-channel microfluidization using natural emulsifiers: Saponins, phospholipids, proteins, and polysaccharides. *Food Hydrocolloids, 61*, 703–711.

Baloch, Q. B., Maher, S., Shah, S. N., Sheeraz, M., Iqbal, N., & Raza, H. (2022). Revitalization of tourism and hospitality sector: Preempting pandemics through lessons learned. *Environmental Science and Pollution Research, 29*(55), 83099–83111. https://doi.org/10.1007/s11356-022-21392-7

Baraero-Era, J. S., & Del Rosario, J. (2020). Examining tourism resilience practices as basis for a post-Covid 19 recovery in the Philippines. *ASEAN journal on Hospitality and Tourism, 18*(1), 1–12. https://doi.org/10.5614/ajht.2020.18.1.01

Bertalanffy, L. V. (1968). General systems theory as integrating factor in contemporary science. *Akten des XIV. Internationalen Kongresses für Philosophie, 2*, 335–340.

Bhandari, L. (2022). Approaches to local community participation in tourism development. *Journal of Tourism & Adventure, 5*(1), 44–61.

Bonfanti, A., Vigolo, V., & Yfantidou, G. (2021). The impact of the Covid-19 pandemic on customer experience design: The hotel managers' perspective. *International Journal of Hospitality Management, 94*, 102871. https://doi.org/10.1016/j.ijhm.2021.102871

Canton, H. (2021a). United Nations conference on trade and development—UNCTAD. In *The Europa directory of international organizations 2021* (pp. 172–176). Routledge.

Canton, H. (2021b). Organisation for economic co-operation and development—OECD. In *The Europa directory of international organizations 2021* (pp. 677–687). Routledge.

Cao, Y., Jian, F., Wang, J., Yu, Y., Song, W., Yisimayi, A., ... & Xie, X. S. (2023). Imprinted SARS-CoV-2 humoral immunity induces convergent Omicron RBD evolution. *Nature, 614*(7948), 521–529.

Chang, P. K. (2014). *Separation of flow*. Elsevier.

Clements, P., Lachenbruch, P., Siebold, J., White, B., Weiner, S., Martin, R., Weinstein, A., Weisman, M., Mayes, M., & Collier, D. (1995). Inter and intra-observer variability of total skin thickness score (modified Rodnan TSS) in systemic sclerosis. *The Journal of Rheumatology*, *22*(7), 1281–1285.

Costa, C. P., Moreira, J. N., Lobo, J. M. S., & Silva, A. C. (2021). Intranasal delivery of nanostructured lipid carriers, solid lipid nanoparticles and nanoemulsions: A current overview of in vivo studies. *Acta Pharmaceutica Sinica B*, *11*(4), 925–940.

Crabolu, G., Font, X., & Miller, G. (2023). The hidden power of sustainable tourism indicator schemes: Have we been measuring their effectiveness all wrong? *Journal of Travel Research*, 00472875231195736.

Department of Budget and Management. (2023). *DBM: Govt allots P17.7B to support tourism infrastructure development*. Department of Budget and Management. https://www.dbm.gov.ph/index.php/management-2/322-dbm-govt-allots-p17-7b-to-support-tourism-infrastructure-development

Deparment of Tourism. (2021). *DOT inks tourism "CARES for TRAVEL" MSME loan program with SB Corp*. Department of Tourism. https://beta.tourism.gov.ph/news_and_updates/dot-inks-tourism-cares-for-travel-msme-loan-program-with-sb-corp/

Dirgantara, A. R., Zallesa, S., & Baswarani, D. T. (2023). Tourist comfort and safety study case study: Ogan Komering Ilir District. *Journey: Journal of Tourismpreneurship, Culinary, Hospitality, Convention and Event Management*, *6*(2), 1–12. https://doi.org/10.46837/journey.v6i2.159

Gaki, E., Kostopoulou, S., Parisi, E., & Lagos, D. (2016). The evaluation of tourism satisfaction in island destinations: The case of the Ionian Islands of Greece. In *56th Congress of the European Regional Science Association: Cities & Regions: Smart, Sustainable, Inclusive?* 23–26 August 2016, Vienna, Austria (pp. 23–26).

Gilmore, A. (2017). Quality and quantity in tourism. *Journal of Hotel & Business Management*, *6*(1), 164.

Glyptou, K. (2022). Tourism sustainability indicators. *Encyclopedia of tourism management and marketing*. Edward Elgar Publishing.

Gretter, A., Ciolli, M., & Scolozzi, R. (2018). Governing mountain landscapes collectively: Local responses to emerging challenges within a systems thinking perspective. *Landscape Research*, *43*(8), 1117–1130. https://doi.org/10.1080/01426397.2018.1503239

Gul, H. (2023). The revival of local tourism. *Imarat Institute Policy Studies*. https://iips.com.pk/the-revival-of-local-tourism/

Gutierrez, E. L. M., Rivera, J. P. R., Roxas, F. M. Y., & Zamora, M. L. (2020). Rebooting Philippine tourism from the COVID19 pandemic. *The World Financial Review*, 28–32. https://worldfinancialreview.com/rebooting-philippinetourism-from-the-covid-19-pandemic/

Harris, A. (2014). Word of mouth in tourism: Reflections and directions. In *CAUTHE 2014: Tourism and hospitality in the contemporary world: Trends, changes and complexity* (pp. 261–272). https://search.informit.org/doi/abs/10.3316/INFORMIT.393780398205406

Hu, H., & Li, C. (2023). Smart tourism products and services design based on user experience under the background of big data. *Soft Computing*, *27*(17), 12711–12724. https://doi.org/10.1007/s00500-023-08851-0

Hu, F., Teichert, T., Deng, S., Liu, Y., & Zhou, G. (2021). Dealing with pandemics: An investigation of the effects of COVID-19 on customers' evaluations of hospitality services. *Tourism Management, 85*, 104320. https://doi.org/10.1016/j.tourman.2021.104320

Huete-Alcocer, N., López-Ruiz, V. R., & Grigorescu, A. (2019). Measurement of satisfaction in sustainable tourism: A cultural heritage site in Spain. *Sustainability, 11*(23), 6774.

Hui, T. K., Wan, D., & Ho, A. (2007). Tourists' satisfaction, recommendation and revisiting Singapore. *Tourism Management, 28*(4), 965–975. https://doi.org/10.1016/j.tourman.2006.08.008

Husain, S. A., Dube, G., Morris, H., Fernandez, H., Chang, J. H., Paget, K., Sritharan, S., Patel, S., Pawliczak, O., Boehler, M., Tsapepas, D., Crew, R. J., Cohen, D. J., & Mohan, S. (2020). Early outcomes of outpatient management of kidney transplant recipients with coronavirus disease 2019. *Clinical Journal of the American Society of Nephrology, 15*(8), 1174–1178.

Inghels, D. (2020). *Introduction to modeling sustainable development in business processes: Theory and case studies*. Springer Nature.

International Monetary Fund. (2020). *Policy responses to COVID19*. International Monetary Fund. https://www.imf.org/en/Topics/imf-and-covid19/Policy-Responses-to-COVID-19

Jaaron, A. A. M., Pham, D. T., & Cogonon, M. E. (2021). Systems thinking to facilitate "double loop" learning in tourism industry: A COVID-19 response strategy. *Journal of Sustainable Tourism, 31*(4), 1–19. https://doi.org/10.1080/09669582.2021.1948554

Javier, A. B., & Elazigue, D. B. (2011, March). Opportunities and challenges in tourism development roles of local government units in the Philippines [Paper Presentation]. In *Third annual conference of the academic network of development studies in Asia*. Nagoya University.

Jiang, R. D., Liu, M. Q., Chen, Y., Shan, C., Zhou, Y. W., Shen, X. R., Li, Q., Zhang, L., Zhu, Y., Si, H. R., Wang, Q., Min, J., Wang, X., Zhang, W., Li, B., Zang, H. J., Baric, R. S., Zhou, P., Yang, X. L., & Shi, Z. L. (2020). Pathogenesis of SARS-CoV-2 in transgenic mice expressing human angiotensin-converting enzyme 2. *Cell, 182*(1), 50–58.

Jiang, Y., Ritchie, B. W., & Verreynne, M. L. (2023). Building dynamic capabilities in tourism organisations for disaster management: Enablers and barriers. *Journal of Sustainable Tourism, 31*(4), 971–996. https://doi.org/10.1080/09669582.2021.1900204

Jong, A. (2022). *7 tips to ensure customer health and safety*. Good Tourism Institute. https://goodtourism.institute.com/library/ensure-customer-health-safety/

Kaushal, V., & Srivastava, S. (2021). Hospitality and tourism industry amid COVID-19 pandemic: Perspectives on challenges and learnings from India. *International Journal of Hospitality Management, 92*, 102707.

Krittayaruangroj, K., Suriyankietkaew, S., & Hallinger, P. (2023). Research on sustainability in community-based tourism: A bibliometric review and future directions. *Asia Pacific Journal of Tourism Research, 28*(9), 1031–1051. https://doi.org/10.1080/10941665.2023.2276477

Kumar, A., Luthra, S., Mangla, S. K., & Kazançoğlu, Y. (2020). COVID-19 impact on sustainable production and operations management. *Sustainable Operations and Computers*, *1*, 1–7. https://doi.org/10.1016/j.susoc.2020.06.001

Kuznetsova, J. A. (2023). Development of the Philippine tourism economy in the post-COVID period. *Servis v Rossii i za rubezhom (Services in Russia and Abroad)*, *17*(2), 87–95. https://doi.org/10.5281/zenodo.8111152

Lannon, C. (2012). Causal loop construction. *The Basics*. https://thesystemsthinker.com/causal-loop-construction-the-basics/

Lei, T., Wang, J. Y., & Pei, J. (2014). Design, synthesis, and structure–property relationships of isoindigo-based conjugated polymers. *Accounts of Chemical Research*, *47*(4), 1117–1126.

Li, P., Zhou, Y., & Huang, S. (2023). Role of information technology in the development of e-tourism marketing: A contextual suggestion. *Economic Analysis and Policy*, *78*, 307–318. https://doi.org/10.1016/j.eap.2023.03.010

Mai, T., & Smith, C. (2015). Addressing the threats to tourism sustainability using systems thinking: A case study of Cat Ba island, Vietnam. *JournCat Ba Islandable Tourism*, *23*(10), 1504–1528. https://doi.org/10.1080/09669582.2015.1045514

Manulife Holdings Berhad. (2019). Annual Report 2019. https://www.manulife.com.my/content/dam/insurance/my/documents/aboutus/financial-statement/Manulife-Annual-Report-2019.pdf

Mei, X. Y., Arcodia, C., & Ruhanen, L. (2010). A national government's tourism innovation initiatives: A review of tourism development policies in Norway [Conference Session]. In *The 21st Council for Australian University Tourism and Hospitality Education Annual Conference (CAUTHE 2011)*, Sydney.

Moaveni, G. (2014). *Determinants of profitability in tourism industry: Evidence from Turkey*. Doctoral dissertation, Eastern Mediterranean University (EMU)-Doğu Akdeniz Üniversitesi (DAÜ). http://hdl.handle.net/11129/1321

Ng, S., Wu, P., Nishiura, H., Ip, D. K., Lee, E. S., & Cowling, B. J. (2011). An analysis of national target groups for monovalent 2009 pandemic influenza vaccine and trivalent seasonal influenza vaccines in 2009–10 and 2010–11. *BMC Infectious Diseases*, *11*, 1–11. https://doi.org/10.1186/1471-2334-11-230

Nguyen, N. C., & Bosch, O. J. (2013). A systems thinking approach to identify leverage points for sustainability: A case study in the Cat Ba Biosphere Reserve, Vietnam. *Systems Research and Behavioral Science*, *30*(2), 104–115.

Osewe, P. L. (2020). *Rebooting the economy: Restoring travel and tourism in the COVID-19 era*. Asian Development Bank. https://blogs.adb.org/blog/rebooting-the-economy-restoring-travel-and-tourismin-covid-19-era

Paasovaara, R., Luomala, H. T., Pohjanheimo, T., & Sandell, M. (2012). Understanding consumers' brand-induced food taste perception: A comparison of 'brand familiarity' – And 'consumer value – Brand symbolism (in)congruity' – Accounts. *Journal of Consumer Behaviour*, *11*(1), 11–20. https://doi.org/10.1002/cb.356

Republic of the Philippines. (2009). *The Official Gazette*. National Printing Office. https://www.officialgazette.gov.ph/

Ritchie, B. W., & Jiang, Y. (2019). A review of research on tourism risk, crisis and disaster management: Launching the annals of tourism research curated collection on tourism risk, crisis and disaster management. *Annals of Tourism Research*, *79*, 102–812. https://doi.org/10.1016/j.annals.2019.102812

Ropret, R. (2014). Injuries in skiing and snowboarding: epidemiology and risk factors as a basis for prevention measures. *Physical Culture* [*Fizicka Kultura*], *68*(2).

Roxas, F. M. Y., Rivera, J. P. R., & Gutierrez, E. L. M. (2018). Framework for creating sustainable tourism using systems thinking. *Current Issues in Tourism*, *23*(3), 280–296. https://doi.org/10.1080/13683500.2018.1534805

Roxas, F. M. Y., Rivera, J. P. R., & Gutierrez, E. L. M. (2021). Bootstrapping tourism post-COVID-19: A systems thinking approach. *Tourism and Hospitality Research*, *22*(1), 86–101. https://doi.org/10.1177/14673584211038859

Roy, D. (2021). The revival of the tourism industry from the Covid-19 pandemic disaster. *International Journal of Tourism & Hospitality Reviews*, *8*(2), 1–7. https://doi.org/10.18510/ijthr.2021.821

Sánchez-Sánchez, M. D., Pablos-Heredero, C. D., & Montes-Botella, J. L. (2021). The internal demand of cultural tourism: Understanding satisfaction and fidelity to destination in Spain through a non-linear structural model. *Sustainability*, *13*(23), 13487. https://doi.org/10.3390/su132313487

Schiuma, G., Carlucci, D., & Sole, F. (2012). Applying a systems thinking framework to assess knowledge assets dynamics for business performance improvement. *Expert Systems with Applications*, *39*(9), 8044–8050. https://doi.org/10.1016/j.eswa.2012.01.139

Schwartz, G., Hugounenq, R., & Clements, B. J. (1995). *Government subsidies: Concepts, international trends, and reform options*. International Monetary Fund. https://www.elibrary.imf.org/view/journals/001/1995/091/article-A001-en.xml

Seabra, C., & Bhatt, K. (2022). Tourism sustainability and COVID-19 pandemic: Is there a positive side? *Sustainability*, *14*(14), 8723. https://doi.org/10.3390/su14148723

Sharma, G. D., Thomas, A., & Paul, J. (2021). Reviving tourism industry post-COVID-19: A resilience-based framework. *Tourism Management Perspectives*, *37*, 100786. https://doi.org/10.1016/j.tmp.2020.100786

Sigala, M. (2020). Tourism and COVID-19: Impacts and implications for advancing and resetting industry and research. *Journal of Business Research*, *117*, 312–321. https://doi.org/10.1016/j.jbusres.2020.06.015

Solomon, M. R. (2002). *Consumer behavior*. Prentice-Hall. https://www.statista.com/statistics/1053908/philippines-number-of-foreign-visitor-tourist-arrivals/

Statista. (2022). Philippines: Foreign tourist arrivals 2022 | Statista. https://www.statista.com/statistics/1053908/philippines-number-of-foreign-visitor-tourist-arrivals/

Steiner, C. (2006). Tourism, poverty reduction and the political economy: Egyptian perspectives on tourism's economic benefits in a semi-rentier state. *Tourism and Hospitality Planning & Development*, *3*(3), 161–177. https://doi.org/10.1080/14790530601132286

Stevens, P., Knutson, B., & Patton, M. (1995). Dineserv: A tool for measuring service quality in restaurants. *Cornell Hotel and Restaurant Administration Quarterly*, *36*, 56–60. https://doi.org/10.1177/001088049503600226

Stone, M. T., & Nyaupane, G. P. (2018). Protected areas, wildlife-based community tourism and community livelihoods dynamics: Spiraling up and down of community capitals. *Journal of Sustainable Tourism*, *26*(2), 307–324.

Tamim, S. I., & Bostwick, J. B. (2020). A dynamic analysis of the Rayleigh–Taylor instability in soft solids. *Extreme Mechanics Letters*, *40*, 100940.

Tam, J. L. M. (2008). Brand familiarity: Its effects on satisfaction evaluations. *Journal of Services Marketing*, *22*(1), 3–12. https://doi.org/10.1108/08876040810851914

The World Bank. (2022, May 30). *Supporting the Philippines' COVID-19 emergency response*. The World Bank. https://www.worldbank.org/en/results/2022/05/30/supporting-ph-covid-19-emergency-response

Torabi, Z. A., Shalbafian, A. A., Allam, Z., Ghaderi, Z., Murgante, B., & Khavarian-Garmsir, A. R. (2022). Enhancing memorable experiences, tourist satisfaction, and revisit intention through smart tourism technologies. *Sustainability*, *14*(5), 2721. https://doi.org/10.3390/su14052721

Upadhaya, S., Tiwari, S., Poudyal, B., Godar Chhetri, S., & Dhungana, N. (2022). Local people's perception of the impacts and importance of ecotourism in Central Nepal. *PLoS One*, *17*(5), e0268637. https://doi.org/10.1371/journal.pone.0268637

Van Heuvelen, E. (2023). *Subsidy wars* (pp. 54–55). Finance & Development.

Van Mai, T., & Maani, K. E. (2009). *Systems thinking for sustainable tourism in the Cat Ba biosphere reserve of Viet Nam*.

Walker, P.A., Greiner, R., McDonald, D., & Lyne, V. (1998). The tourism futures simulator: A systems thinking approach. *Environmental Modelling & Software*, *14*(1), 59–67.

Williams, M. L., Burnap, P., & Sloan, L. (2017). Towards an ethical framework for publishing Twitter data in social research: Taking into account users' views, online context and algorithmic estimation. *Sociology*, *51*(6), 1149–1168.

Won, H., Huang, J., Opland, C. K., Hartl, C. L., & Geschwind, D. H. (2019). Human evolved regulatory elements modulate genes involved in cortical expansion and neurodevelopmental disease susceptibility. *Nature Communications*, *10*(1), 2396.

Won, J. Y., Lee, Y. R., Cho, M. H., Kim, Y. T., & Heo, B. Y. (2022). Impact of government intervention in response to coronavirus disease 2019. *International Journal of Environmental Research and Public Health*, *19*(23), 16070. https://doi.org/10.3390/ijerph192316070

WTTC. (2021). *#SafeTravels: Global protocols & stamp for the new normal*. World Travel & Tourism Council. https://wttc.org/initiatives/crisis-preparedness-management-recovery/safetravels-global-protocols-stamp

Yehia, Y. (2019). The importance of tourism on economies and businesses. *Global Edge Official Portal*. https://globaledge.msu.edu/blog/post/55748/the-importance-of-tourism-on-economies-a

Zenker, S., & Kock, F. (2020). The coronavirus pandemic – A critical discussion of a tourism research agenda. *Tourism Management*, *81*, 104–164. https://doi.org/10.1016/j.tourman.2020.104164

Zhou, X., Mei, R., & Song, L. (2023). Is the government's direct subsidy to COVID-19 effective? *Heliyon*, *9*(5), e15707. https://doi.org/10.1016/j.heliyon.2023.e15707

Zimik, S., Barman, A., & Dey, R. (2022). Re-designing tourists' satisfaction models through review triangulations. *Manager – The British Journal of Administrative Management*, *58*(154). https://ssrn.com/abstract=4393436

Chapter 7

Active Transport for Low-Carbon Tourism

Jonna C. Baquillas[a], *Marie Danielle V. Guillen*[b] *and Edieser DL. Dela Santa*[b]

[a]De La Salle University, Philippines
[b]University of the Philippines-Diliman, Philippines

Abstract

As the tourism industry recovers from the devastating effects of the global pandemic, meeting the targets of the Sustainable Development Goals (SDGs) remains to be a global "deadline" where tourism is seen as a major contributor. While disruptions to business-as-usual practices such as COVID-19 present unprecedented challenges, they can also provide opportunities for strategic innovation to change behavior toward sustainable tourism experiences. Active transport for low-carbon tourism such as walking or cycling tours have risen in popularity in recent years, and especially postpandemic, as they provide opportunities for a more personalized experience while health and safety protocols can still be implemented. They also present health benefits for the individuals while contributing to environmental sustainability and climate mitigation strategies of the tourism industry. This book chapter presents cases of various forms of tourism activities that use active transport, focusing on walking tours and cycling tours. Various companies offering tours under these modes are discussed and presented. These two modes promote authentic cultural and heritage tourism experiences through the local experts that provide the services.

Keywords: Sustainable mobility; sustainable tourism; low-carbon mobility; walking tour; cycling tour

Introduction

The general objective of this chapter is to describe forms of active transport for low-carbon tourism. The description is set against an explanation of post-pandemic tourism trends where it is noted that demand is slowly recovering across

Revisiting Sustainable Tourism in the Philippines, 135–148

doi:10.1108/978-1-83753-678-820241007

the regions. This chapter points to calls for more low-carbon approaches. Active transport is thus identified as a mechanism to manage tourism growth and to achieve sustainability. A brief description of the components of active transport for tourism as well as illustrative cases from across the region sketch the range of tourism products and experiences that may revolve around the concept. This chapter ends with a discussion of some challenges and recommendations for the future.

The global pandemic disrupted the economic growth of tourism. As this sector is regarded an important contributor in promoting local economic development (Baquillas & Gozun, 2022), disruptions led to significant reduction in business performance of tourism companies, increase in unemployment rates, and closure of some businesses due to a decline in demand for long-distance travel and tourism services, restrictions on tourism-related activities, and physical distancing requirements (Jin et al., 2023). This also negatively impacted the tourism supply chain, especially because tourism services are heavily reliant on in-person commercial activities, thereby reducing consumption of tourism products (Lee et al., 2024). This resulted to low occupancy rates, business closures (temporary or permanent), additional costs of disinfection protocols and use of personal protective equipment (PPE), and cancellation of reservations (González-Torres et al., 2021).

Postpandemic Tourism Trends

Tourism has gradually but steadily reopened since 2021 (Brilhante & Rocha, 2023). International tourism is expected to restore its prepandemic levels by 2024, pointing at a 2% increase based on 2019 prepandemic levels as forecasted by the United Nations World Trade Organization (UNWTO, 2024). This was a good recovery from a dip in international tourist arrivals in 2020 (409 million), to a slight improvement in 2021 (448 million) (UNWTO, 2022), and a more promising growth in 2022 (900 million) (UNWTO, 2023a, 2023b) and 2023 (estimated at 1.3 billion) (UNWTO, 2024). This progress also paved the way for job recovery, as around 18.2 million jobs were made available again in the tourism sector (Brilhante & Rocha, 2023).

Some changes in tourist behavior have been observed upon the resumption of tourism activities. For example, visiting national parks experienced systematic growths driven by the need for outdoor activities and the desire to experience nature as the toll of the pandemic set in (Jodłowski et al., 2023). While national parks are important to nature preservation, they also are rooted on significant cultural heritage and social value. The protection given to national parks means that sustainability efforts in preserving biodiversity and in ensuring that tourism does not create a negative impact on its existence need to be considered. This suggests that there needs to be a balance between providing a satisfying experience to tourists while minimizing the environmental impacts of their presence (Jodłowski et al., 2023).

Given the renewed enthusiasm of tourists to travel again, mass tourism remains a concern. The unsustainable destination visits to popular locations pose threat to livelihoods and environment and cause congestions (Cooke et al., 2022). Overtourism, when not controlled or managed, equates to diminished quality of life for the locals and reduced tourism experience for the tourists (Ghaderi et al., 2022). A call for tourism "degrowth" has been proposed by a few academics as a "new normal" scheme, where carrying capacity limits and social/cultural considerations are respected. Reform in the form of degrowth is emphasized, rather than going back to how things were prepandemic (Butcher, 2021).

In terms of transport choices, nonpreference for use of public transportation has been observed in a coastal town in Catalan, Spain. While public transport provides accessibility and a more sustainable means of mobility, the pandemic affected its use and caused a decline in its utilization (Delclòs-Alió et al., 2022). Supporting this, the use of active transport in tourism has seen a dramatic increase, as tourists see biking or walking as a COVID-19-secure means of going around (Ito & Cheer, 2022).

Active transport is highly encouraged as it promotes good physical health while reducing emissions from motorized transport (Kim & Hall, 2022a). Active transport for low-carbon tourism that consists of walking or biking tours has seen a dramatic increase in recent years and more so postpandemic as it promotes personalized cultural and tourism experiences while managing physical distancing protocols (Ito & Cheer, 2022). This aspect is further developed in the next section.

Demand for Low-Carbon Tourism Strategies

The SDGs emphasize the need to rethink strategies in the way things are done that involve various stakeholders (Chiu et al., 2020), and the tourism sector is no exception. As a main contributor in sustainable development, this sector has an important role in promoting social and economic progress while minimizing environmental impacts (Giurca et al., 2022).

The use of active transport was heightened during the pandemic, as the share of trips taken by bicycles rose substantially, impacting travel behavior. Recreational cycling grew in the United States, especially during weekends, while transit-oriented cities in Germany shifted to biking instead of taking public transportation (Buehler & Pucher, 2021). In Korea, a 37.4% increase in sales of bicycles was observed during the pandemic. The main drivers for shifting to cycling for leisure or work include personal health, public health, attitude, and subjective norms (Kim & Hall, 2022b).

In the Philippines, local government units strategically implemented bicycle lanes in major thoroughfares to accommodate and protect those who wish to use this nonmotorized transport to travel. An estimated 36% or 10 million households have at least one member who uses a bicycle for mobility based on the Social Weather Stations survey conducted in March 2023 (Cervantes, 2023). This increased from 7.3 million households (29%) in 2022 and 6.2 million households (24%) in 2021. The pandemic opened up a window of opportunity for active transport to be recognized and mainstreamed as the government approved the allocation of a budget worth PHP 1.1B (23 M USD) in September 2020 for a 500-km bike lane network in three metropolitan cities in the Philippines, namely Metro Manila, Metro Cebu, and Metro Davao (Sunio & Babiano, 2022).The increase may be partly due to the filing of the Senate Bill 1698 or the Bicycle Act, which allows for the setup of infrastructures that support bicycle use including dedicated lanes and parking areas (Senate Bill Aims to Create Nationwide Bike Lanes, Safety Laws, n.d.).

Thailand has created cycling-related activities in celebration of Queen and Rama IX, which has been popular since 2015. More people started using bicycles for fitness and tourism (Sann et al., 2023). Meanwhile, Vietnam has robustly promoted the use of bicycles and has even attracted western tourists who come for cycling holidays (Matteucci & Tiller, 2023).

Active Transport and Low-Carbon Tourism

With tourism demand on track to recovery from the pandemic, approaches to low-carbon tourism have never been more important. This is because the global tourism industry is a major contributor to the growth of greenhouse gases (GHGs), including carbon dioxide (CO_2), methane, nitrous oxides, hydrofluorocarbons, perfluorocarbons, sulfur hexafluoride, and other short-lived GHGs (Gossling et al., 2013). Gössling (2009) finds that the expansion of emissions by the industry as a whole outperforms improvements in reducing emissions per tourist. Breaking this down at the sector level, Gossling et al. (2013) share reports made in 2005 that pinpoint tourism transport, accommodation, and activities as contributing about 5% to global emissions by humans of CO_2. CO_2 emissions are associated with transport, with aviation contributing approximately 40% of tourism's overall carbon footprint, followed by car transport (32%), and accommodation (21%). More updated statistics shared by Becken (2023) has air travel pushing its contribution to 74% of tourism transport emissions, with much of the emissions generated in developed countries. Moreover, based on estimates by UNWTO and United Nations Environment Programme (UNEP), it was feared that continued growth on business-as-usual basis would mean that CO_2 emissions from tourism would grow 135% by 2035 vis-à-vis the statistics of 2005, about 3.059Gt, a figure quite close to another estimate by the World Economic Forum of 3.164 Gt by 2035.

As a result, the question of how destination growth is to be managed has become undoubtedly important. Scholars such as Dwyer (2023) have noted how this question is embedded in such growth management strategies as green growth, inclusive growth, and even the SDG 2030 agenda. A range of specific policies have been proposed including: energy-efficiency gains in buildings, the use of biofuels, carbon off-setting, GHG disclosure, educational programs and awareness building, energy-efficiency measures, incentives, investments, information technology to facilitate low-carbon choices, low-carbon mobility, mobility management, public transport, development of renewable energies, strategic assessment of emissions to identify mitigation options, sewage treatment, spatial planning, solid waste management, taxation, and the enforced use of low-carbon technology Gossling et al. (2013). It is asserted that even small gains in the implementation of these policies would mean steps toward attaining low-carbon growth and development.

The identification of low-carbon mobility could not have been more appropriate, as it ultimately relates to walking and cycling, the "pre-eminent forms of low carbon travel" (Weston & Mota, 2012, p. 1). They comprise the key elements of active transport or active travel, which refers to any mode of transport where physical activity is incorporated (Perez et al., 2017). The bicycle in particular has been acknowledged as an example of a "convivial technology" or "convivial tool," that Jess (2023, p. 139) explains, from the words of Illich (1973), as a technology that is simple to understand, communally owned, democratically developed, and autonomously operated and repaired based on the active involvement of its users, thus, a means for degrowth.

Indeed, there is extensive literature on the multiple benefits and effects of cycling in managing tourism growth and in terms of social, economic, and environmental sustainability. As a form of slow tourism, cycle tourism allows tourists to experience a territory and all that it represents – culture and traditions, history, landscape, gastronomy, agriculture, and many others – at a speed defined on a human scale (Gazzola et al., 2018). Increased active travel is found to enhance well-being and work–life balance from individual physical activity (Herman & Larouche, 2021; Perez et al., 2017) as well as from better air quality brought about by the reduced pollution due to cycle use (Ciascai et al., 2022). Zafri et al. (2021) emphasize how issues of public health in cities are better addressed using active transport mode.

Growth in active transport tourism is also associated with economic benefits. Cycle tourism leads to the creation of new job descriptions such as mobility manager in order to meet the requirements of cycle tourists (Gazzola et al., 2018). Investments in related infrastructure such as routes and trails, complementary products and services including bike sharing and rental systems, e-bikes, and public service transportation of bicycles, as well as food and beverage outlets generate direct and indirect economic gains for destinations as diverse as cities to remote mountain regions (Lee & Huang, 2012; Mundet et al., 2022; Taylor et al., 2022).

The environment is likewise seen to benefit from increased active transport tourism. Strongly emphasizing the compatibility of active transport modes with greenways, Di Ruocca et al. (2020) raise the capacity of bicycles in reducing carbon emissions in the transport network. The construction of green roads is assessed to lead to a considerable lessening of GHG emissions (Ngo et al., 2018) as well as substantial increase in biking and walking (Frank et al., 2021). A further benefit is an increase in demand for the creation and protection of green spaces as part of much larger initiatives to develop active transport tourism infrastructure (Lasco, 2023).

Indeed, the need to develop infrastructure is one of the challenges. The pandemic has exposed how much of the streets are designed for motorized vehicles particularly cars, hence, the need to make them safer for active travelers through changes in road configuration, construction of walkways, bike lanes, trails, and similar spaces (Palafox, 2022). An essential infrastructure component is a system of routes that links sites to one another and integrates bicycles with other modes of public transport (Lee & Huang, 2012).

Investigating the long-term resilience of trail systems in New Zealand, Taylor et al. (2022) discuss not just physical links but also managing the connectivity within and between internal system components as well as those external to the system including the formal and informal network between tourism service providers and the creation of links with external bodies through marketing activities or the staging of cycling events. They likewise emphasize keeping track of changes in the preferences of visitors, offering services that meet requirements, and forming participatory governance structures.

The importance of government action and the public's response toward low-carbon mobility was highlighted during the pandemic. Researchers have noted how government agencies globally invested in active transport tourism infrastructures and facilities to encourage people, and the response has been positive. Zafri et al. (2021) report how in Bangladesh, like in many other countries, walking increased, and bicycles sales zoomed upwards. Their study also demonstrates the importance of infrastructure for active transport, with respondents expected to travel more by cycling where supportive infrastructure were available. In Korea, Kim and Hall (2022a), investigating the influence of personal and public health and smart applications on biking behavior, reinforce findings regarding the personal and public health motivations when engaging in active travel. An important part of infrastructure was protected bike lanes. In the Philippines, the government was trumpeting the length of bike lanes designed and built during the pandemic. While there were safety concerns regarding a plan to share the use of the lanes with motorcycles on some roads in Metro Manila (Cruz, 2023), these have been allayed by local government interventions (Villamente, 2024) and the continuing expansion of the active transport program as laid out in the national transport policy and the Philippine Development Plan 2023–2028.

Illustrative Cases: Active Transport's Role in Promoting Tourism Experiences

As highlighted in the earlier section, active transport involves human-powered modes of transportation like walking, cycling, and hiking. These activities play a significant role in promoting tourism experiences. Such human-powered modes of transportation are not just low-carbon but a zero-carbon mode of tourism activity from a purely active transport definition.

It is thus important to understand the components of active transport tourism in the context of transportation system and tourism sector. Kim and Hall (2022b) highlighted in their study how active transport contributes to both health benefit and emission reduction and how walking and biking for tourism is different from other leisure activities. It also highlighted why sustainable destination planning is important.

The components of active transport tourism are no different from the components of transportation system, but the mode, a major component, becomes more personal – the mode, in this case, is the tourist (the person or the walker or cyclist). The appeal is on the tourist's behavior and deeper appreciation of the place or it can be as Kim et al. (2018) noted the appeal to prosocial behavior or also known as responsible tourism. Second, the built environment or the infrastructure – it has been established that automobile dependence is positively associated with single use, low-density land development, and disconnected street networks (Frank et al., 2006). In this regard, better street design for walking and cycling is a primary infrastructure component needed for active transport tourism. And finally, there is the system of operations including governance and management of active transport tourism. Governance is an important part of this as it provides overall policy guidance with regards to the built environment, especially walkways and cycling lanes. In the case of management, active transport tour operators or organizations are at the forefront of providing the experience that tourists expect like heritage or culinary tourism in the case of walking or cycling tours.

Active transport in the tourism industry can be broadly categorized into two: Walking Tour and Cycling Tour. These two can be further subclassified according to types of tours.

(1) Walking Tours are activities that utilize human power and can be further described into different types such as:

- Hiking Tours: This is illustrated by hikers ascending through a picturesque mountain trail or hill, surrounded by breathtaking landscapes. This enables tourists to access remote and unexplored areas, providing them with unparalleled experiences and sense of accomplishment. This kind of tours is often packaged as adventure trail tours. Companies such as TrailAdventours in the Philippines offer different kind of hikes depending on fitness level and budget packages.[1]
- Cultural Walking Tours: This tourism activity is usually guided by local experts who share stories of historical sites, traditional markets, and hidden gems. This kind of walking tour promotes a slow and immersive experience, allowing visitors to absorb the culture and heritage of a destination, thus getting some attention from tourists. In Thailand, an organization called the Bangkok Walking Tours actively promotes walking tours in the area.[2] It was founded by two bloggers, a Bangkok native and a Spaniard from Barcelona. They offer premade tours ("Pick A Tour"), customized tours ("Make A Tour"), and joiner tours that start at their office in Khaosan Road. Their activities focus on walking, letting

[1]Trail Adventours is one of the local organizations in the Philippines that offers guides for hiking as part of their tourism adventure packages.

[2]Bangkok Walking Tours is just a project started by bloggers Wut Suthirachartkul and Pol Comaposada under the company Mundo Nomada Travel Company Limited in 2012.

you see Bangkok in its authenticity, while the guides offer insightful knowledge about the city. This company offers different options and even includes the use of public transport as part of the alternatives (About – Bangkok Walking Tours – Get Around Like a Local, 2016).
Napong Tao Rugkhapan (2023) noted in his study how walking tours – a particular kind of linear tourism – are used to weave disparate sites into a singular narrative that frames Charoenkrung in Thailand as a multicultural district with creative potential. In particular, it sheds light on the role of the tour guide in charting this entangled narrative of multiculturalism and creativity. A series of thematic walking tours called Creative District Tour is designed to introduce the public to different aspects of Charoenkrung and its surroundings. Based on his study, linear tourism tightens two things: space and narrative, which are particularly useful for a spread-out district like the case of Charoenkrung. Reducing GHG may not be its target, but it helps in promoting the importance of walking for low-carbon transport.

- Walking Food/Culinary Tour: Walking food tours are also gaining traction in Asia. Active transport enhances culinary tourism by allowing visitors to savor regional flavors and specialties while enjoying the scenic routes. In the Philippines, a walking food tour called the Big Binondo Food Wok by Old Manila Walks has received the highest rating of five in TripAdvisor.[3] It is a 2.5-hour tour by Ivan Man Dy that takes the tourists to a variety of food stalls, eateries, and restaurants and is considered as one of the best things to do in Manila according to some bloggers like willyforfood (Renée, 2023).[4]
- Eco-parks or National Parks and Wildlife Safaris: These activities allow tourists to explore a nature reserve on foot, observing animals in its natural habitat. Active transport supports sustainable tourism practices by minimizing the ecological impact and promoting a responsible approach to wildlife encounters. In Singapore, they are also known as Green Tours. There are tour operators like Monster Day Tours that offer intimate walking trail tours, bringing tourists around the heart of the city, encompassing Raffles Place, Marina Bay, and the historic Singapore River.[5] The tour ends at the verdant oasis of Gardens by the Bay, a 101-hectare nature park.

(2) Cycling Tours are the kind of tours aided by the use of bicycle.
Road cycle tourism is a type of cycling which, by virtue of its characteristics, can help in adjusting tourist demand in destinations with very rigidly marked tourist seasons. This type of activity using bicycles can also help in moving tourists away from urban centers toward the more rural or less developed areas and help promote the inclusion of new tourist agents

---

[3]Old Manila Walk is one of the leading tour outfits in the Philippines that runs heritage-themed walks around Manila's historic districts.
[4]Willyforfood – one of the award winning blogs for travel and gastronomy.
[5]Monster Day Tours – an award-winning tour organization in Singapore.

Mundet et al. (2023). Active transport is also a means to manage traffic in tourism destinations, while catalyzing the mode shift from motorized transport to walking and cycling (Thanh et al., 2020).

Cycling tours also come in different forms:

- Nature Cycling: This involves cycling through scenic nature trails, surrounded by lush greenery. Active transport allows visitors to immerse themselves in the natural beauty of a destination, fostering a deeper connection with the environment. In Singapore, one can explore the gardens and the surrounding urban landscape via bike. This is being offered by Let's Go Bike Singapore, a private tour operator. This company has been given the Best Tour Experience by the Singapore Tourism Award from 2020–2022 (A, 2023).[6]
- City Cycling Tours: This includes guided cycling tour through the historic streets of a city, with tourists pedaling through cultural landmarks, vibrant neighborhoods, and iconic attractions. Active transport provides an intimate and unique way to explore urban destinations, allowing tourists to engage with the local culture. This kind of cycling tour has been introduced by Bambike, initially a socio-ecological enterprise that sells handmade bamboo bikes but now also provides environmentally friendly bike tours in Intramuros, a historic walled city inside Manila.[7] Its products range from all-terrain bikes such as the Luntian and Victoria, the sleek, bolt-together Bolt Series, and "Bambino" bikes for kids. They offer bike tours known as the Bambike Intramuros Experience. It is also interesting to note that from Intramuros, the company has started a rental shop in Guimaras province, an idyllic community known for its use of renewable energy (Bambike Ecotours, n.d.). Bambike has also created bicycle fleets for hotels in Siem Reap, Cambodia. Cambodia is another country known for its use of pedal power for touring Angkor Wat, which is considered as a "classic tour" by most.

The importance of active transport cannot be underestimated. Walking is the very first activity that humans do and the most underrated mode to appreciate one's culture and surroundings. Cycling is also known as an advanced form of walking since one can control the way they move via bicycle. The above illustrations show how active transport has been utilized to get to know and promote a city or a destination.

---

[6]Let's Go Singapore – an award-winning tour operator offering a wide range of unique tours led by experienced local experiential tours in Singapore.

[7]Bambike – a socio-ecological enterprise based in the Philippines that handmakes bamboo bicycles with fair-trade labor and sustainable building practices and offers bike tour services in Intramuros, Manila.

Conclusion, Recommendations, and Policy Implications

This chapter was concerned with describing active travel as a means to achieve more sustainable outcomes, namely low-carbon tourism, amid lessons from the pandemic and the excesses of overtourism before that. As the chapter shows, a range of tourism products and experiences revolving around walking and cycling can be developed. Their use has grown over the years and certainly over the pandemic years, as market preferences have shifted, the industry slowly adopted, and governments have ramped up investments in active transportation.

As tourism destinations reopen postpandemic, a more sustainable approach to tourism activities has been reimagined. Reinvigorating the economy has opened opportunities for sustainable strategies in the tourism sector. Active transport for low-carbon tourism brings in a different way of experiencing culture and heritage of a destination, while at the same time addresses the negative environmental impacts usually associated with tourism activities. Walking and cycling tours offer a more intimate glimpse into the local life and culture, as tourists enjoy exploring on foot or on a bike. Cases in the Philippines, Thailand, and Singapore show the willingness of tourists to undergo unique experiences associated with active transport. The role of government as exemplified in the case of Japan and the Philippines in supporting infrastructure for active transport tourism is also an important component for the shift to happen.

In transitioning to low-carbon tourism, it would be necessary to acknowledge some of the challenges. After all, as some scholars and industry reports (Dwyer, 2023; Gupta & Woetzel, 2022) have noted, it has been extremely challenging and painful to reduce emissions. These include addressing knowledge gaps in tourism growth regions particularly on impacts, adaptation measures, and their effectiveness against emissions (Scott et al., 2023), which is very important if sustainability strategies like active transport were to be adopted more broadly. In addition, the tourism industry in Asia is still dominated by conventional modes of production and consumption, reinforcing observations that within the region, the industry is not deeply involved in sustainable practices (Tolkach et al., 2016), even when a major environmental rehabilitation program is in their backyard (Legaspi & Dela Santa, 2022). All these may lead to overtourism and/or promote unsustainable practices particularly as operators wish to recoup investments lost from the pandemic.

Thus, transformative policies need to be incorporated in existing tourism sector strategies in order to support initiatives such as the use of active transport. As the tourism sector slowly but steadily recovers from the ill effects of the pandemic, this presents the perfect opportunity to rethink tourism management and integrate policies to promote active transport for low-carbon tourism.

Dedication

This manuscript is wholeheartedly dedicated to the scholar who put us together on his passing – Prof. Brian C. Gozun, whose light continue to shine upon us and inspire us to do scientific work.

References

A, A. (2023, August 23). *About us. Let's go bike Singapore.* https://www.letsgobikesingapore.com/about-us/

About – Bangkok Walking Tours – Get around like a local. (2016, May 4). *Bangkok walking tours – Get around like a local.* https://bangkokwalkingtours.com/about-usabout/

Bamboo bike tours in Manila, Philippines | Bambike Ecotours. (n.d.). Bambike. https://www.bambike.com/ecotours

Baquillas, J. C., & Gozun, B. C. (2022). Strolling between Shanties: Tourists' perceptions and experiences of Manila's slums. In *Perspectives on Asian tourism* (pp. 59–78). https://doi.org/10.1007/978-981-19-4013-2_4

Becken, S. (2023). *Tourism transformation or business-as-usual? Many hands make light work 6th Philippine research conference on tourism and hospitality, October 23–24.* University of the Philippines Asian Institute of Tourism.

Brilhante, M. F., & Rocha, M. L. (2023). COVID-19 pre-pandemic tourism forecasts and post-pandemic signs of recovery assessment for Portugal. *Research in Globalization*, 100167. https://doi.org/10.1016/j.resglo.2023.100167

Buehler, R., & Pucher, J. (2021). COVID-19 impacts on cycling, 2019–2020. *Transport Reviews*, *41*(4), 393–400. https://doi.org/10.1080/01441647.2021.1914900

Butcher, J. (2021). Covid-19, tourism and the advocacy of degrowth. *Tourism Recreation Research*, *48*(5), 633–642. https://doi.org/10.1080/02508281.2021.1953306

Cervantes, F. (2023). *SWS: 1 out of 3 Filipino families uses bicycles.* Philippine News Agency. https://www.pna.gov.ph/articles/1208953

Chiu, A. S., Aviso, K. B., Baquillas, J., & Tan, R. R. (2020). Can disruptive events trigger transitions towards sustainable consumption? *Cleaner and Responsible Consumption*, *1*, 100001. https://doi.org/10.1016/j.clrc.2020.100001

Ciascai, O. R., Dezsi, S., & Rus, K. A. (2022). Cycling tourism: A literature review to assess implications, multiple impacts, vulnerabilities, and future perspectives. *Sustainability*, *14*(8983), 1–18. https://doi.org/10.3390/su1415898

Cooke, P., Nunes, S., Oliva, S., & Lazzeretti, L. (2022). Open innovation, soft branding and green influencers: Critiquing 'fast fashion' and 'overtourism'. *Journal of Open Innovation: Technology, Market, and Complexity*, *8*(1), 52. https://doi.org/10.3390/joitmc8010052

Cruz, J. P. (2023, August 22). Shared bike-motorcycle lane on EDSA: Concerns raised over MMDA plans. https://www.rappler.com/philippines/metro-manila/safety-concerns-mmda-plans-shared-bicycle-motorcycle-lane-edsa-group-raises/

Delclòs-Alió, X., Gutiérrez, A., Miravet, D., Tomàs-Porres, J., Vich, G., & Clavé, S. A. (2022). Impact of the COVID-19 pandemic on tourist public transportation use and on its determinants: Evidence from a Catalan coastal destination. *Tourism Management Perspectives*, *44*, 101003. https://doi.org/10.1016/j.tmp.2022.101003

Di Ruocca, G., Iglesias, L. P., Blandon, B., & Melella, R. (2020). Low-carbon tourism – Technical economic and management project of a greenway, for enhancing inner areas of the Cilento National Park, Italy. *Sustainability*, *12*(23), 10012. https://doi.org/10.3390/su122310012

Dwyer, L. (2023). Tourism degrowth: Painful but necessary. *Sustainability*, *15*(20), 14676. https://doi.org/10.3390/su152014676

Frank, L. D., Hong, A., & Ngo, V. D. (2021). Build it and they will cycle: Causal evidence from the downtown Vancouver Comox Greenway. *Transport Policy*, *105*, 1–11.

Frank, L. D., Sallis, J. F., Conway, T. L., Chapman, J. E., Saelens, B. E., & Bachman, W. (2006). Many pathways from land use to health: Associations between neighborhood walkability and active transportation, body mass index, and air quality. *Journal of the American Planning Association*, *72*(1), 75–87. https://doi.org/10.1080/01944360608976725

Gazzola, P., Pavione, E., Grechi, D., & Ossola, P. (2018). Cycle tourism as a driver for the sustainable development of little-known or remote territories: The experience of the Apennine regions of Northern Italy. *Sustainability*, *10*, 1863. https://doi.org/10.3390/su10061863

Ghaderi, Z., Hall, C. M., & Ryan, C. (2022). Overtourism, residents and Iranian rural villages: Voices from a developing country. *Journal of Outdoor Recreation and Tourism*, *37*, 100487. https://doi.org/10.1016/j.jort.2022.100487

Giurca, A., Béfort, N., & Taylor, A. (2022). Exploring transformative policy imaginaries for a sustainable post-COVID society. *Journal of Cleaner Production*, *344*, 131053. https://doi.org/10.1016/j.jclepro.2022.131053

González-Torres, T., Rodríguez-Sánchez, J., & Pelechano-Barahona, E. (2021). Managing relationships in the tourism supply chain to overcome epidemic outbreaks: The case of COVID-19 and the hospitality industry in Spain. *International Journal of Hospitality Management*, *92*, 102733. https://doi.org/10.1016/j.ijhm.2020.102733

Gössling, S. (2009). Carbon neutral destinations: A conceptual analysis. *Journal of Sustainable Tourism*, *17*(1), 17–37. https://doi.org/10.1080/09669580802276018

Gossling, S., Scott, D., & Hall, C. M. (2013). Challenges of tourism in a low-carbon economy. *WIRE's Clim Change*, *4*, 525–538. https://doi.org/10.1002/wcc.243

Gupta, R., & Woetzel, J. (2022). Asia's net-zero transition: Opportunity and risk amid climate action. https://www.mckinsey.com/featured-insights/future-of-asia/asias-net-zero-transition-opportunity-and-risk-amid-climate-action. Accessed on January 24, 2024.

Herman, K. M., & Larouche, R. (2021). Active commuting to work or school: Associations with subjective well-being and work-life balance. *Journal of Transport & Health*, *22*, 101118. https://doi.org/10.1016/j.jth.2021.101118

Illich, I. (1973). Re-tooling society. *Philippine Studies*, *21*(1/2), 125–185. http://www.jstor.org/stable/42632212

Ito, E., & Cheer, J. (2022). Everyday cycling mobilities and tourist behaviour: The emergence of cycle tourism as COVID-secure activity. *Journal of Responsible Tourism Management*, *2*(1), 10–17. https://doi.org/10.47263/jrtm.02-01-02

Jess, L. M. (2023). Degrowth and the slow travel movement: Opportunity for engagement or consumer fad. In *Tvergastein debates in post-development and degrowth* (Vol. 2, pp. 134–150). https://politicalecologynetwork.org/wp-content/uploads/2023/01/04853-degrowthvolume2.pdf#page=69

Jin, C., Cong, Z., Dan, Z., & Zhang, T. (2023). COVID-19, CSR, and performance of listed tourism companies. *Finance Research Letters*, *57*, 104217. https://doi.org/10.1016/j.frl.2023.104217

Jodłowski, M., Kruczek, Z., Szromek, A., & Gmyrek, K. (2023). Tourists' attitudes towards visitor management and restrictions in the national parks in the Carpathian mountains. *Studia Periegetica*, *42*(2).

Kim, M. J., & Hall, C. M. (2021). Is tourist walkability and well-being different? *Current Issues in Tourism*, *26*(2), 171–176. https://doi.org/10.1080/13683500.2021.2017409

Kim, M. J., & Hall, C. M. (2022a). The influence of personal and public health and smart applications on biking behavior in South Korea. *Journal of Consumer Behavior*, *22*, 382–395. https://doi.org/10.1002/cb.2076

Kim, M. J., & Hall, C. M. (2022b). Does active transport create a win-win situation for environmental and human health? The moderating effect of leisure and tourism activity. *Journal of Hospitality and Tourism Management*, 487–498. https://doi.org/10.1016/j.jhtm.2022.08.007

Kim, M. J., Park, J. Y., Reisinger, Y., & Lee, C. K. (2018). Predicting responsible tourist behavior: Exploring pro-social behavior and perceptions of responsible tourism. *International Journal of Tourism and Hospitality Research*, *32*(4), 5–20.

Lasco, G. (2023, September 15). *Wanted: Walkable, bikeable cities*. Philippine Daily Inquirer.

Lee, C., & Huang, H. (2012). The attractiveness of Taiwan as a bicycle tourism destination: A supply-side approach. *Asia Pacific Journal of Tourism Research*, *19*(3), 273–299. https://doi.org/10.1080/10941665.2012.739190

Lee, L., Wang, Y., Zhang, L., Ping, L., Zuo, J., & Zhang, H. (2024). Impact of COVID-19 on the economic loss and resource conservation of China's tourism industry from the supply chain perspective. *Cities*, *144*, 104633. https://doi.org/10.1016/j.cities.2023.104633

Legaspi, G. F. A., & Dela Santa, E. (2022). Partnerships towards sustainability: The revival of Boracay's wetlands. In A. S. Balasingam & Y. Ma (Eds.), *Asian tourism sustainability* (pp. 19–38). Springer.

Matteucci, X., & Tiller, T. R. (2023). Package cycle tourists' relationship to time and pace. *Tourism and Hospitality Research*, *23*(3), 332–343. https://doi.org/10.1177/14673584221102701

Mundet, L., Marin, J., & Figueroa, A. (2022). How to develop a road cycling tourism destination: Girona as a case study. *Journal of Outdoor Recreation and Tourism*, *39*, 100566.

Ngo, V. D., Frank, L. D., & Bickerstaff, K. (2018). Effects of new urban greenways on transportation energy use and greenhouse gas emissions: A longitudinal study from Vancouver, Canada. *Transportation Research Part D: Transport and Environment*, *62*, 715–725.

Palafox, Jr., F. A. (2022, April 25). *Walkable, bikable and livable cities*. Philippine Daily Inquirer.

Perez, K., Olabarria, M., Rojas-Rueda, D., Santamaria-Rubio, E., Borrell, C., & Nieuwenhuijsen, M. (2017). The health and economic benefits of active transport policies in Barcelona. *Journal of Transport & Health*, *4*, 316–324.

Rugkhapan, N. T. (2023). Linear tourism, multiculturalism, creative district: The case of Charoenkrung creative district in Thailand. In *Annals of tourism research* (Vol. 102). https://doi.org/10.1016/j.annals.2023.103626

Renée, J. (2023, May 22). *Binondo food Wok with Ivan Man Dy: What to do in Manila.* Will Fly for Food. https://www.willflyforfood.net/what-to-do-in-manila-binondo-food-wok-with-ivan-man-dy/

Sann, R., Lai, P., & Chen, C. (2023). Crisis adaptation in a Thai community-based tourism setting during the COVID-19 pandemic: A qualitative phenomenological approach. *Sustainability, 15*(1), 340. https://doi.org/10.3390/su15010340

Scott, D., Hall, C. M., Rushton, B., & Gossling, S. (2023). A review of the IPCC Sixth Assessment and implications for tourism development and sectoral climate action. *Journal of Sustainable Tourism.* https://doi.org/10.1080/09669582.2023.2195597

Senate bill aims to create nationwide bike lanes, safety laws. (n.d.). Philippine News Agency. https://www.pna.gov.ph/articles/1193735

Sunio, V., & Babiano, I.-M. (2022). Pandemics as 'windows of opportunity': Transitioning towards more sustainable and resilient transport systems. *Transport Policy, 116*, 175–187. https://doi.org/10.1016/j.tranpol.2021.12.004

Taylor, P., Frost, W., & Frost, J. (2022). Building resilient local tourism systems: Insights from New Zealand's premier cycling attractions. *Asia Pacific Journal of Tourism Research, 27*(5), 457–472.

Thanh, T. T., Van Dao, M. D., & Ly, H. B. (2020). Analysis of tourist travel behavior and recommendation for active transport encouragement strategies, the case of Hue City lecture notes in civil. *Engineering, 54*, 1049–1055.

Tolkach, D., Chon, K. K., & Xiao, H. (2016). Asia Pacific tourism trends: Is the future ours to see? *Asia Pacific Journal of Tourism Research, 21*(10), 1071–1084.

UNWTO. (2022). *World tourism barometer* (p. 20). United Nations World Tourism Barometer.

UNWTO. (2023a, January 17). *Tourism set to return to pre-pandemic levels in some regions in 2023.* UNWTO. https://www.unwto.org/news/tourism-set-to-return-to-pre-pandemic-levels-in-some-regions-in-2023. Accessed on November 30, 2023.

UNWTO. (2023b, May 9). *Tourism on track for full recovery as new data shows strong start to 2023.* UNWTO. https://www.unwto.org/news/tourism-on-track-for-full-recovery-as-new-data-shows-strong-start-to-2023. Accessed on November 29, 2023.

UNWTO. (2024, January 19). *International tourism to reach pre-pandemic levels in 2024.* UNWTO. https://www.unwto.org/news/international-tourism-to-reach-pre-pandemic-levels-in-2024#:~:text=According%20to%20the%20first%20UNWTO, estimated%201.3%20billion%20international%20arrivals. Accessed on January 27, 2024.

Villamente, J. (2024, January 8). *QC bike lanes barriers will be in the form of plant boxes.* Daily Tribune.

Weston, R., & Mota, J. C. (2012). Low carbon tourism travel: Cycling, walking and trails. *Tourism Planning & Development, 9*(1), 1–3. https://doi.org/10.1080/21568316.2012.658168

Zafri, N. M., Khan, A., Jamal, S., & Alam, B. M. (2021). Impacts of the Covid-19 pandemic on active travel mode choice in Bangladesh: A study from the perspective of sustainability and new normal situation. *Sustainability, 13*, 6975. https://doi.org/10.3390/su13126975

Chapter 8

Women and Sustainable Tourism: Experiences of Environmental Protection and Conservation in the Philippines

Eylla Laire M. Gutierrez

Asian Institute of Management – Dr. Andrew L. Tan Center for Tourism, Philippines

Abstract

Environmental protection and conservation have been closely linked with the promotion of sustainable tourism across destinations. This is underpinned by the assumption that the environment and tourism have an interdependent relationship – where tourism activities need to support the preservation and conservation of environmental areas so that such economic activities can be sustained. That is a balance must be struck between the two. Within these two areas, women have been recognized to play a crucial role – in tourism, women were seen to dominate the industry's workforce globally; in environmental efforts, women were believed to play a key role in supporting such initiatives across societies. In both areas, women's efforts, despite being crucial to the achievement of sustainability, are often perceived as small and thus overlooked. Despite these, little has been said about their experiences in promoting sustainable tourism. To contribute to discourses promoting the idea of "thinking small," this chapter looks at women's tourism-related environmental efforts that are often seen as small and overlooked, alongside their experiences in being involved in these initiatives.

Keywords: Women; sustainability; tourism; environment; Philippines

Introduction

Sustainable tourism is readily conceptualized in terms of its "people, planet, and profit" components – where tourism contributes to social, environmental, and

Revisiting Sustainable Tourism in the Philippines, 149–158

Published under exclusive licence by Emerald Publishing Limited
doi:10.1108/978-1-83753-678-820241008

economic aspects of development. Among these three aspects, sustainability has been widely associated with efforts relating to environmental protection and conservation (Purvis et al., 2019). This is underpinned by an understanding that tourism's sustained development is largely dependent on the protection and conservation of the environmental areas in which it thrives (Telfer & Sharpley, 2015). This is particularly apparent for nature-based tourism offerings (i.e., ecotourism, marine tourism, and beach tourism, among others). Thus, efforts to promote tourism activities that support environmental areas have been widely advocated among stakeholders and key players in the industry. Within these initiatives, women were believed to play a pivotal role.

In tourism, women were recognized to comprise the majority of the tourism workforce. According to the United Nations World Tourism Organization (UNWTO) (2020), women make up 59% of the tourism workforce globally. Despite this, only 23% of women occupy leadership roles, while most occupy low-skilled and lower paid positions across the industry (Mouriki, 2023). This reflects the reality that women's experiences working in the industry vary (Gutierrez & Vafadari, 2023) – where tourism can be a tool to promote their empowerment and where tourism exacerbates gender stereotypes and inequalities (Gutierrez, 2024). While women are believed to play a leading role in promoting sustainable tourism, their potential remains curtailed by a range of issues relating to their gender and their social standing in communities and societies (Gutierrez & Vafadari, 2023).

On the other hand, in the field of environmental conservation and protection, women were seen to play a crucial, if not a leading role, in promoting environmental protection and conservation (Agarwal, 1992). Theoretically, this belief is supported by Ecofeminism (Agarwal, 1992; Howell, 2004) which argues for the presence of an inextricable and natural link between women and the environment.[1] In the developmental field, the role of women in environmental initiatives has been widely advocated since the Rio Declaration in 1992.[2] More recently, pronouncements on women's substantial position in environmental issues have been coupled with recognitions that they are more at risk (than men) of environment-related issues and disasters (UN Women, 2023). Like women's complex dynamics within the tourism industry, their standing in the environmental field proves equally challenging.

Within these realms of tourism, environment, and development agendas, three assumptions have been widely accepted: first, tourism and the environment are closely linked with each other; second, women play a crucial role in tourism development and the protection and conservation of environmental areas; third, women play an important role in the promotion of sustainable tourism. Yet, many scholars have pointed out the danger in accepting them as general truths (Jackson, 1993; Rao, 1991; Resurrección, 2013). That is, context is crucial in better understanding the varied experiences and interactions between women and the environment.

---

[1]That is, the "domination of women and the domination of nature" (Howell, 2004, p. 231).
[2]The Rio Declaration (1992) stated: "Women have a vital role in environmental management and development. Their full participation is therefore essential to achieving sustainable development."

To contribute to discourses promoting "thinking small," this chapter looks at women's tourism-related environmental efforts that are often seen as small and overlooked, alongside their experiences in being involved in these initiatives. To do so, the following research question is posed: "How are women involved in tourism-related environmental conservation and protection activities?" Taking off from the works of Gutierrez (2024), Jumawan-Dadang (2015), and Ondiba and Matsui (2021), the study examines the multiple ways in which women experience and negotiate their relationship with tourism and the environment. By looking at the experiences of women celebrated for their environmental contributions in the Philippines, the study aims to showcase how women are involved and motivated to promote environmental protection and conservation. Insights support existing works that highlight the diversity of women's experiences in being involved with tourism and the environment (Gutierrez, 2024; Jumawan-Dadang, 2015; Ondiba & Matsui, 2021). The study also supports the findings suggesting that women employ a variety of ways to negotiate their roles in societies through their engagements with the environment. The insights from the study offer perspectives on how more women can be engaged, involved, and mobilized to contribute to sustainable tourism (Mosse & Lewis, 2006) even through small but meaningful and impactful ways.

Women, Tourism, and Environmental Protection in the Philippines

Women and Tourism in the Philippines

Women have played a crucial role in the development of the Philippines's tourism industry. For years now, women's presence in the industry is increasingly becoming more apparent albeit the roles they play remain complex and varied (Gutierrez, 2024; Gutierrez & Vafadari, 2023). The limited yet insightful literature on women reveals this. For one, the early work of Chant (1996) has focused on women's presence in sex tourism industries in the Philippines which mirrored society's perspective on women as an economic commodity. Carlos et al. (2023) found that the majority of the informal workers in Boracay are women. The informality of their work proves that they often engage in work that does not offer economic security and safety given their educational background and socioeconomic statuses. Women's presence in informal entrepreneurial activities in tourism areas was also highlighted by Gumba (2020). Arguably, women's engagement in microenterprises in tourism areas is influenced by their lack of access to credit given that they are still perceived as secondary income earners. In extending this finding, Gutierrez (2024) found that social aspects of women's engagement in tourism remain complex as an imbalance is observed between the perception of their households on their work and the perceptions of their communities on their work. Arguably, women's engagement in economic activities, specifically tourism, remains a prospect frowned upon by their own families. Dulnuan and Mondiguing (2000), on the other hand, found that women's work in tourism often exacerbates existing gender stereotypes – where women are engaged in unpaid jobs related to domestic or care work.

Despite these challenges, however, findings have also revealed that women benefited from their engagement in tourism work. Pleno (2006), for example, found that women's work in tourism enhanced social and psychological well-being as they were found to lead happier lives through their involvement in the industry. The same is observed by Gutierrez (2024) where women were found to have improved psychological empowerment as they were able to enhance their self-confidence while also developing skills that they were not present before they engaged with tourism (i.e., managerial and communication skills, among others). Dulnuan and Mondiguing (2000) and Gumba (2020) also cited that through their engagement in tourism, women were able to contribute to their household income. Carlos et al. (2023) also supported these findings by pointing out that tourism offers an opportunity for women to participate in economic activities given the flexibility it offers to them. That is, alongside women's domestic and household responsibilities, they were able to participate in income-generating activities that also improved their statuses. Across these studies, it has become apparent that women, despite the challenges they face, actively seek ways to improve their psychological, social, political, and economic positions within communities and societies through tourism (Carlos et al., 2023; Gumba, 2020; Gutierrez, 2024; Gutierrez & Vafadari, 2023; Pleno, 2006).

Women and Environmental Protection and Conservation in the Philippines

Historically, women were found to be the protectors of the environment in the Philippines. Before colonization, women from indigenous communities were known preservers and protectors of the environment (Gabriel et al., 2020). Women's knowledge of the conservation and protection of environmental areas is also well-recognized (Gabriel & Mangahas, 2017). Given the vital role of women as care providers of households, they are believed to have a proclivity to have closer connections with the environment – where women source food and water and alternative medicine, among other needs from the environment areas to which they belong (Naganag, 2014). For these reasons, women deem the environment as a "life support system" (Guiriba, 2013, p. 130) where their lives depend on the environment's health and sustainability.

At present, their close relationship and engagement with the environment remain unparalleled, especially in the case of women in indigenous communities in the country. Their experiences in their engagement in environmental work, similar to tourism, also varied. Naganag (2014), for example, found that the indigenous women of Kalinga continued to safeguard traditions of preserving potable water. Gabriel and Mangahas (2017) noted that women in Nueva Ecija are designated persons to enforce and control environmental activities in areas declared sacred and protected areas. Gabriel et al. (2020) also found that as keepers of traditions, women are designated to perform rituals that are essential in commencing any activities in environmental areas. On the other hand, women were also found to be engaged in crucial yet unpaid and often neglected environmental work. For example, Gabriel et al. (2020) found that women act as overseers of efforts to protect forests from illegal hunters and in managing the *kaingin* system of cultivation to ensure the conservation of environmental areas.[3] The same role is observed by Macusi et al. (2023) in suggesting that women are engaged in often neglected jobs (i.e., watchers and overseers of fish wardens and coastal areas) in marine and fisheries protection.

Women's Experiences in Tourism-Related Environmental Work in the Philippines

In addressing the aims and objectives of this chapter, an exploratory case study approach to investigating women's experiences in tourism-related environmental work will be conducted. To do so, qualitative secondary data will be analyzed including relevant online newspaper and magazine articles, reports, and journal articles, among others. More specifically, the cases examine the roles and experiences of women engaged in tourism-related environmental work. The cases explored herein are women-led efforts that emphasize women's role as leaders and as workers to promote environmental protection and conservation in their areas.

Women in the "No-to-174 Campaign" in Dumaguete City, Philippines

Dumaguete City is the capital of the province of Negros Oriental, Philippines. Known as the City of Gentle People, Dumaguete City is the most populous city of the province despite it being the smallest in terms of land area.[4] As a coastal city, Dumaguete is favored by a rich marine ecosystem that supports its tourism activities (Manahan, 2021). Some of the popular tourist attractions include Apo Island, a top diving and snorkeling spot in the country whose marine life is protected by a small community; Pulangbato Falls Valencia, a unique water attraction adorned with red rocks located at the Red River Valley; Balinsasayao Twin Lakes Natural Park, located in a small crater lake with a view of the mountains, among others (Larena, n.d.). More recently, however, a Smart City reclamation plan posed a challenge to the sustained environmental conservation initiatives of the city's locality. The Smart City project, a public–private partnership project, entails a 174-hectare reclamation plan to create a digitally advanced city expansion within 5 years of its signing (The Manila Times, 2021). The plan was supposed to be signed in July 2021 but was put to a halt after relentless public pressure from its locals which became a famous movement, the "No-to-174 campaign." Contributing to the success of this campaign is "crowd-sourced activism" (Bello, 2022, p. 1) – where community members including

[3]An indigenous practice of intermittent clearing and burning of forest areas for growing and cultivating crops.

[4]Dumaguete is nicknamed "The City of Gentle People" given friendliness and warmth locals provide to its visitors.

fisherfolks, academe, scientists, church groups, and the youth came together to show their opposition toward the plan. At the helm of this campaign are the significant contributions of women, who from various professions have contributed to the movement in their little ways:

One of the key players in the movement was a lawyer who mobilized volunteers and the community to examine the legality of the reclamation plan. Through the work she conducted, she was able to conclude that the proponent private company, EM Cuerpo Inc. (EMCI), does not have the financial capacity to take on such a huge project amounting to about USD 400 million. From there, further examinations were made to prove the legality of the development project. Meanwhile, a teacher and writer also collaborated with artists and youth groups in producing a publication, *Lutong Baybay*, which features local seafood recipes of women from Dumaguete's six fishing communities that will be heavily affected by the reclamation plan (Ecarma, 2022).[5] Other women have dedicated their platforms to raising environmental awareness. One of the most striking ways was utilized by an ophthalmologist who became "the self-appointed social media influencer" (Bello, 2022) of the movement dedicated her social media account to creating information videos to raise concerns about the plan. She even took bikini photos to accompany informative captions that aimed to pique the audience's attention regarding the environment (Bello, 2022).

Amid the success of the movement to stop the development project, the community members of Dumaguete remained vigilant for any other emerging attempts to disturb their environment. To this, a continued effort to raise environmental awareness for the entire community of Dumaguete is deemed necessary by the people themselves. These women, along with other members of the community, continue to work in their small ways to contribute to a more solid ground in protecting their environment – a crux to the continuity of them becoming a "City of Gentle People."

Women in Ecotourism Conservation

Located in Aloguinsan, Cebu on the Visayan Island of the Philippines stands a multi-awarded and internationally recognized community-based tourism initiative, the Bojo River Eco-Cultural Tour. Before being recognized as one of UNWTO's Best Tourism Villages in 2021, Bojo also formed part of the Top 100 Global Sustainable Destinations in 2016, 2017, and 2018 (San Juan, 2021). The success of the tour is primarily owed to the local community that operates it, the Bojo Aloguinsan Ecotourism Association (BAETAS) which has been historically dominated by women and has been women-led. Since its establishment in 2009, BAETAS has been managed and operated by housewives who often had free time as their husbands worked as fisherfolks. Created to act as an alternative source of

[5]A local zine featuring local recipes of women coming from six fishing communities in Dumaguete City. The story illustrates how people living close to the sea will face issues in the face of unmanaged development (i.e., reclamation plan).

income and a measure to eliminate threats to the environment (i.e., an attempted mining construction), BAETAS developed the Bojo River Cruise with environmental and cultural preservation at its core. Part and parcel of these efforts include environmental education for visitors and community members, regular coastal and river clean-ups, and wildlife protection measures, among others. Beyond its environmental efforts, the river cruise has been known for financing community development initiatives and projects that aim to improve the well-being of the community and empower women and children (UNWTO, 2020).

It is usually the woman in the community who manages the overall operations of the cruise alongside the environmental conservation efforts that they are doing. Women in this case can be seen as empowered individuals who are active also in influencing political decisions as they create voting blocks and encourage local politicians to share their platforms with the community. Within the community, both men and women take whatever job they can do with little to no regard for traditionally assigned jobs for men and women – women as fisherfolks and tour guides, men as cooks, and men and women as traditional performers (Gutierrez & Vafadari, 2023). As earlier suggested by Gutierrez (2019), local community participation has been considered one of the success factors of BAETAS.

Conclusion

This chapter sheds light on the prospects of "thinking small" by exploring the potential of those traditionally deemed marginalized and underrepresented in the efforts promoting sustainability such as women. As illustrated in this study, women can become active partners in promoting sustainable initiatives including environmental preservation and conservation. To do this, however, there is a need to recognize the realities that they face – both constraints and opportunities that influence their capacity to support sustainability.

As suggested in this chapter, in both tourism and environmental work, several studies have noted the influential role of gender dynamics in their involvement and participation. As suggested by Gutierrez and Vafadari (2023), women's roles in tourism are largely influenced by gender divisions of labor – where women are engaged in occupations that mirror their household work. Like the experiences of women involved in tourism, most of the environmental work women are involved in is underpaid and even unpaid (Gabriel et al., 2020; Macusi et al., 2023). With women seen as secondary income earners and primarily in charge of domestic work, both women in tourism and environmental work take on multiple burdens by engaging in both productive and reproductive work. For example, women engaged in communal environmental work are also expected to take on their usual domestic work (i.e., preparing and cooking meals for their families, and washing of dishes) on top of their work (Acosta et al., 2016; Naganag, 2014). The same observation was made by Gutierrez (2024) in noting the additional burdens women have to take on when domestic work is not shared with men. Due to women's tendency to be engaged in informal work, both for tourism and environmental work, they become more susceptible to a variety of hazards. Chandra et al. (2017), Gabriel and Mangahas (2017), and Van Impe (2000), for example, noted that under cases of climate events, women were found to experience increased discrimination (i.e., loss of land rights), food insecurity, resource poverty, gender-based violence, and forced migration among others. The same observation is made by Carlos et al. (2023) in noting that women engaged in informal work are often vulnerable to harassment from both tourists and members of the industry. Their positions in these fields are further worsened by the lack of social and institutional systems to safeguard and protect them (Carlos et al., 2023; Gabriel et al., 2020). Within the larger schemes, women's occupation of decision-making and leadership positions are crucial in improving their statuses (Guiriba, 2013) in both fields.

References

Acosta, L. A., Eugenio, E. A., Macandog, P. B. M., Magcale-Macandog, D. B., & Lin, E. K. (2016). Loss and damage from typhoon-induced floods and landslides in the Philippines: Community perceptions on climate impacts and adaptation options. *International Journal of Global Warming*, *9*(1), 33e65. https://doi.org/10.1504/IJGW.2016.074307

Agarwal, B. (1992). The gender and environment debate: Lessons from India. *Feminist Studies*, *18*(1), 119–158. https://doi.org/10.2307/3178217

Bello, R. P. (2022, April 6). Dumaguete: 'City of gentle people' fights to save coastal life. *Rappler*. https://www.rappler.com/environment/city-gentle-people-fights-save-coastal-life-dumaguete-reclamation/

Carlos, J. C. T., Gutierrez, E. L. M., & Bautista, M. J. D. (2023). *Gendering the informal tourism sector toward inclusive and sustainable growth: The case study of Boracay Island*. (No. DP 2023-35). https://ideas.repec.org/p/phd/dpaper/dp_2023-35.html

Chandra, A., McNamara, K. E., Dargusch, P., Caspe, A. M., & Dalabajan, D. (2017). Gendered vulnerabilities of smallholder farmers to climate change in conflict-prone areas: A case study from Mindanao, Philippines. *Journal of Rural Studies*, *50*, 45–59. https://doi.org/10.1016/j.jrurstud.2016.12.011

Chant, S. (1996). Women's roles in recession and economic restructuring in Mexico and the Philippines. In T. Sinclair (Ed.), *Gender, work and tourism* (pp. 297–327). Routledge.

Dulnuan, J. R., & Mondiguing, R. P. (2000). *Gender and tourism in the Cordillera: A study on how tourism affects the women and men of Banaue, Ifugao*. Master's Thesis, University of the Philippines.

Ecarma, L. (2022, April 30). 'Lutong Baybay': Women of Dumaguete's love letter to the sea. *Rappler*. https://www.rappler.com/life-and-style/food-drinks/lutong-baybay-women-dumaguete-love-letter-sea/

Gabriel, A. G., De Vera, M., & B. Antonio, M. A. (2020). Roles of indigenous women in forest conservation: A comparative analysis of two indigenous communities in the Philippines. *Cogent Social Sciences*, *6*(1), 1–18. https://doi.org/10.1080/23311886.2020.1720564

Gabriel, A. G., & Mangahas, T. L. S. (2017). Indigenous people's contribution to the mitigation of climate variation, their perception, and organizing strategy for sustainable community based forest resources management in Caraballo Mountain, Philippines. *Open Journal of Ecology*, *7*(2), 85–100. https://doi.org/10.4236/oje.2017.72007

Guiriba, G. O. (2013). Ecofeminism: Women's work on environmental conservation in Sorsogon Province, Philippines. *The International Journal of Social Sciences*, *13*(1), 123–133. extension://efaidnbmnnnibpcajpcglclefindmkaj/https:/web.archive.org/web/20180421075748id_/ http://tijoss.com/TIJOSS%2013th%20Volume/MR.%20GLENTON%20O.%20GUIRIBA.pdf

Gumba, B. G. (2020). Entrepreneurial engagement of women in selected tourism areas in the Philippines. *International Journal of Scientific Research in Science and Technology*, *7*(4), 349–359. https://doi.org/10.32628/IJSRST207455

Gutierrez, E. L. M. (2019). Participation in tourism cases on community-based tourism (CBT) in the Philippines. *Ritsumeikan Journal of Asia Pacific Studies*, *37*(1), 23–36. https://www.jstage.jst.go.jp/article/rjaps/37/1/37_23/_article/-char/ja/

Gutierrez, E. L. M. (2024). Women participation, empowerment, and community development in select tourism destinations in the Philippines. In *Routledge handbook on gender in tourism* (pp. 94–109). Routledge.

Gutierrez, E. L. M., & Vafadari, K. (2023). Women in community-involved tourism enterprises: Experiences in the Philippines. *Asia Pacific Viewpoint*, *64*(1), 85–97. https://doi.org/10.1111/apv.12361

Howell, N. R. (2004). Ecofeminism: What one needs to know. *Zygon*, *32*(2), 231–241. https://doi.org/10.1111/0591-2385.861997085

Jackson, C. (1993). Doing what comes naturally? Women and environment in development. *World Development*, *21*(12), 1947–1963. https://doi.org/10.1016/0305-750X(93)90068-K

Jumawan-Dadang, R. (2015). Saving marine life: An empirical assessment of ecofeminist thought in coastal communities. *Philippine Sociological Review*, 61–83. http://www.jstor.org/stable/24717160

Larena, E. (n.d.). Top Dumaguete tourist spots: Diving with turtles in Apo Island & natural hot springs. *Guide to the Philippines*. https://guidetothephilippines.ph/articles/what-to-experience/dumaguete-tourist-spots

Macusi, E. D., Maynawang, I. S., Katikiro, R. E., Suyo, J. G., & Macusi, E. S. (2023). Women lead small-scale fisheries to sustainability in Surigao del Sur, Philippines. *Marine Policy*, *155*, 105768. https://doi.org/10.1016/j.marpol.2023.105768

Manahan, J. (2021, July 22). Dumaguete reclamation to devastate nearby protected areas, marine biodiversity—Conservation group. *ABS-CBN News*. https://news.abs-cbn.com/news/07/22/21/wwf-concerned-dumaguete-proposed-land-reclamation-project

Mosse, D., & Lewis, D. (2006). Theoretical approaches to brokerage and translation in development. In D. Mosse & D. Lewis (Eds.), *Development brokers and translators* (pp. 1–26). Kumarian.

Mouriki, D. (2023, August 25). Empowering women in the tourism industry: Bridging the opportunity gap. https://www.linkedin.com/pulse/empowering-women-tourism-industry-bridging-gap-despina-mouriki/

Naganag, E. M. (2014). The role of indigenous women in forest conservation in upland Kalinga province, Northern Philippines. *International Journal of Advanced Research in Management and Social Sciences*, *3*(6), 75–89. https://www.indianjournals.com/ijor.aspx?target=ijor:ijarmss&volume=3&issue=6&article=008

Ondiba, H. A., & Matsui, K. (2021). Drivers of environmental conservation activities among rural women around the Kakamega forest, Kenya. *Environment, Development and Sustainability*, *23*, 10666–10678.

Pleno, M. J. L. (2006). Ecotourism projects and women's empowerment: A case study in the province of Bohol, Philippines. *Forum of International Development Studies*, *32*, 137–155. https://dl.ndl.go.jp/view/download/digidepo_8380218_po_08.pdf?contentNo=1&alternativeNo=

Purvis, B., Mao, Y., & Robinson, D. (2019). Three pillars of sustainability: In search of conceptual origins. *Sustainability Science*, *14*, 681–695. https://doi.org/10.1007/s11625-018-0627-5

Rao, B. (1991). *Dominant constructions of women and nature in the social science literature*. CES/CNS pamphlet 2. University of California.

Resurrección, B. P. (2013). Persistent women and environment linkages in climate change and sustainable development agendas. *Women's Studies International Forum*, *40*, 33–43. https://doi.org/10.1016/j.wsif.2013.03.011

San Juan, A. D. (2021, December 06). Cebu's Barangay Bojo among 'best tourism villages' in the world. *Manila Bulletin*. https://mb.com.ph/2021/12/06/cebus-barangay-bojo-among-best-tourism-villages-in-the-world/

Telfer, D. J., & Sharpley, R. (2015). *Tourism and development in the developing world*. Routledge.

The Manila Times. (2021, October 5). Dumaguete to build 174-hectare smart city. https://www.manilatimes.net/2021/10/05/business/real-estate-and-property/dumaguete-to-build-174-hectare-smart-city/1817144

UN Women. (2023, April 19). Empowering women, protecting the earth. https://www.unwomen.org/en/news-stories/feature-story/2023/04/empowering-women-protecting-the-earth

United Nations World Tourism Organization (UNWTO). (2020). *Women's empowerment and tourism*. https://www.unwto.org/gender-and-tourism

Van Impe, K. (2000). People for sale: The need for a multidisciplinary approach towards human trafficking. *International Migration*, *38*(1), 113e191. https://doi.org/10.1111/1468-2435.00117

III

Recalibrating Tourism Products and Services Through Customization

11

Recalibrating [illegible] Services Through [illegible]

Chapter 9

Revisiting Resiliency and Sustainability of the Tourism and Built Environment Nexus: Macroeconomic and Microlevel Evidence From the Philippines Toward the "Next Normal"

Luisito C. Abueg[a] *and Iris L. Acejo*[b]

[a]University of the Philippines Los Baños, Philippines
[b]Cardiff University, UK

Abstract

As the Philippines enters into the era of the "next normal," that is, from the pandemic era called the "new normal," we look into how the Philippines' state of tourism and built environment sectors. We also revisit its situation having the "longest lockdown in the world" and look into iterations of socioeconomic models, business platforms, and other elements that lead to the dream of a postpandemic sustainable tourism industry. In this inquiry, we highlight the importance of elements of the built environment, both at the macroeconomic level and at the micro units of business, civil society, and the tourism sector at large. While there has been a substantive discussion on the nexus of the tourism and built environment sectors, little has been devoted to the challenges these intertwined sectors faced during the COVID-19 pandemic. We suggest complementation of practices from the macroscale to the microlevel tourism and built environment sectors, and vice-versa, which will ensure the full complementation of the sectors. These proposals are in full contextualization of the objective toward the postpandemic continued recovery, growth, and sustainability, from the local, national, and the regional economy of Southeast Asia, and also to various parts of the world.

Keywords: Sustainable tourism; built environments; COVID-19 pandemic; "new" normal; cruise ports

Revisiting Sustainable Tourism in the Philippines, 161–179

doi:10.1108/978-1-83753-678-820241009

Introduction: The Philippine Macroeconomic Performance Surrounding the COVID-19 Pandemic

The Philippines – together with the rest of the world – was significantly affected by the unprecedented COVID-19 pandemic. The pandemic has affected almost every aspect of human life, resulting into sudden mobility disruptions causing a global economic recession that was unseen almost a century since the 1918 Spanish Flu and World War 1 (Abueg, 2020a). The pandemic also was coupled with the Ukraine–Russia War, which triggered increases in oil prices and further economic and political uncertainties. These international geopolitical events have also contributed to the postpandemic surge in inflation and exchange rates, with the Philippines having its recorded historical lowest USD-PHP exchange rate of 59 to a dollar in October 2022 (Agcaoili, 2023). Fortunately, exchange rates stabilized at the Philippine Peso (PHP) 55 to USD 1 level in 2023 and in early 2024. The country was notoriously known also for having the "longest lockdown in the world" (Abueg, 2020b), with the lowest gross domestic product growth coupled with highest unemployment and persistently high underemployment rates in the first 2 years of the pandemic. These dismal macroeconomic indicators are indicative of how the country was largely unprepared in mitigating the effects of the pandemic at the very least. Further, the health-induced recession was different in the sense that the learnings of the previous economic crises are not appropriate in the pandemic mitigation efforts toolkit, and that the usual inflation-recession duo did not manifest (Abueg, 2020a, 2020b).

One of the unfortunate sectors that have endured much of the pandemic's negative effects was the real estate and development sector and the tourism sectors (Abueg et al., 2021; Acejo & Abueg, 2022). Both sectors largely depend on services as output, labor as input, and the mobility of consumers in creating demand for the outputs. On the one hand, the Philippine tourism sector reached the eight-million mark in inbound tourism arrivals in 2019 just before the pandemic (Cabalza, 2020). The pandemic has significantly reduced this number, with local tourist arrivals showing a decline (Acejo & Abueg, 2022; Caynila et al., 2022). On the other hand, the real estate development sector has its own share of "series of unfortunate events."[1] Data recorded by the Philippine Statistics Authority (PSA) have indications showing that the pandemic accelerated the real estate cycle (Abueg et al., 2021). Notably, several players in the real estate sector have started to engage in real estate investment trusts (REITs) and becoming more integrated with the sector's portfolio (Abueg et al., 2022). Contrastingly, not all players benefitted in this portfolio, and the enabling law has been in place since 2009: 11 years prior to the pandemic (Fig. 9.1).[2]

---

[1]In discussing the sorry state of the Philippine economy during the COVID-19 pandemic, we are reminded of the book series *A Series of Unfortunate Events* by Lemony Snicket. This metaphor in the title mirrors that of the experiences of both tourism and real estate sectors during the pandemic, and as well the national economy at large.

[2]The Philippine Congress enacted Republic Act no. 9856 that allowed real estate and development companies and corporations to put up REITs.

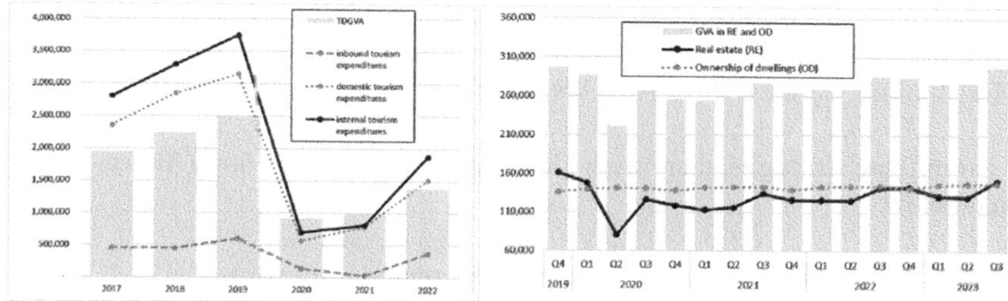

Fig. 9.1. [Left] Total Domestic Gross Value Added of Tourism, Inbound Tourism Receipts, Domestic Tourism Expenditures, and Internal Tourism Expenditures in Current Prices, Units in Million PHP, 2017–2023. [Right] Gross Value Added (GVA) in Real Estate (RE) and Ownership of Dwellings (OD) in Constant 2018 Prices, Units in Million PHP. *Source:* Department of Tourism and PSA.

Given macroeconomic and microlevel data and studies, we also investigate the complementation of the tourism and real estate development sectors and suggest programs and policies that would concordantly enhance this nexus. We also trace and investigate potential channels and mechanisms of growth and sustainability of the tourism sector, from the region's platforms and programs to the Philippine Tourism Development Plan (PTDP), and up to the level of the local tourism sectors. With the backdrop of resiliency, we look at the practices that worked before and during the pandemic and those that needs to be adjusted, improved, and requires innovation. The survey of models before and during the pandemic will provide us with a renewed perspective on how the sectors will be sustainable in the "next normal."

Review of Tourism Programs Before the Pandemic: The Philippines and Southeast Asia

The ten-member bloc of Southeast Asia has been praised by the international community in terms of economic viability and potential (Breene, 2016). More particularly, the region's tourism sector has been one promising source of value added for the region's economy, as well as for its member states. These paved for the ASEAN's programs on tourism, which are also reflected in the region's vision – the ASEAN Economic Community – with general program indicated in the ASEAN Economic Blueprint (Abueg, 2023a). The complementary document on the "Plan of Action for ASEAN Tourism Cooperation" (ASEAN Secretariat, 2018) indicated the following key strategies:

- Marketing the ASEAN region as a single tourist destination with multifaceted attractions and world-class standards and facilities.
- Encouraging tourism investments under a more competitive regime.

- Developing a critical pool of tourism manpower.
- Promoting environmentally sustainable tourism.
- Facilitating seamless intra-ASEAN travel.

From these strategies, there were identified programs and plans (Abueg, 2023a) listed below:

- Cross-border infrastructure to complement regional tourism.
- The single aviation market, open skies policy, and unified visa policy scheme.
- Connectivity and technology.
- Inclusive tourism and economic development.
- Cultural tourism, merchandising, and microenterprises.

Additionally, there were separate ASEAN declarations on cruise tourism and gastronomy tourism were made in 2018 (Abueg, 2023a). These declarations were targeted to increase international demand on the region's tourism products and services. Member states thus align their national and local tourism programs to conform with the general programs of action by the ASEAN.

In the Philippines, such alignment may be seen in the introduction of the "one town, one product" (OTOP) program spearheaded by the Department of Trade and Industry (DTI). The program aims to promote local demand for goods and services, which will help most microenterprises and small and medium-scale firms and companies and currently pushed for institutionalization via national law (Cervantes, 2021). The Department of Tourism (DOT) is also active in branding the Philippine tourism sector through slogan-making and addressing the international assessments and cross-country rankings and metrics of tourism infrastructure, services, and policies.

Along with the importance of physical infrastructure in the promotion of tourism sector, digital infrastructure equally becomes significant as a forefront response to COVID-19 pandemic. As the world's migration to the digital space accelerated, big data are also generated that is essential in decision-making of both the public sector, the business sector, the tourism sector, and the civil society at large. The complementarity of physical and digital infrastructure in the business endeavors of the tourism sector shows one way for the sector's resilience, renewed sustainability, and more opportunities for growth.

Elements of Tourism and Built Environment: Overview

Literature on tourism and built environment shows that the intersectionality between the two sectors is large, with these sectors affecting and influencing each other (Ewing & Cervero, 2001, 2010; UNSD, 2016, 2022). Beyond the usual physical infrastructure of buildings, transportation that complements the natural environment, there is also a recognition of man-made structures and elements such as architecture, culture and history, and digital technology that blends the tourism and built environment sectors (Specht, 2014; UNESCO, 2023). In the Philippines, much of these elements of the built environment are also seen in the tourism portfolio of the country.

One example is the increased appreciation and recognition of natural elements, historical structures, and architecture that contribute to evolving tourism activities in the last decades (Capistrano & Muñoz, 2022; Pilapil-Añasco & Lizada, 2014), which is also embodied in the Philippine Development Plan (PDP) 2023–2028 (NEDA, 2023). In Manila alone, the surviving Art Deco structures during the American period were refurbished and repurposed as the triumvirate of the National Museum (Fine Arts, Natural History, and Anthropology). Restoration of the Metropolitan Manila Theater (or Met) and maintenance of other historical buildings were notable despite the national government's limited budget – thanks to civil society movements and donations. However, recent events of the fire in the National Archives in 2021 (Sarao, 2021) and the Manila Central Post Office in 2023 (Magramo, 2023) and the related fire incident that burned partially the newly restored Met in 2022 (Ong, 2022) purported concerns regarding shortcomings in maintaining such historical structures (Adel, 2023a; Ines, 2023; Maglinong, 2018). These architectural marvels show consistency with the argument in Specht (2014) that architecture also plays a role not only in the spheres of built environment but also the experience that it creates making their locations tourism-relevant places. Recent restoration efforts directed government officials to include these structures to tourism portfolios and as part of tourism feats and the cited structures above as part of the "Manila tourism circuit" (Licsi, 2024). In particular, the planned restoration of the Manila Central Post Office was undertaken by the national government in cooperation with the National Commission for the Culture and the Arts or NCCA, the National Historical Commission of the Philippines or NHCP, and the local government of the City of Manila. This is a part of a long-term and urban development project to transform Pasig River into a tourist destination (Alcazaren, 2024; Boiser et al., 2023; Doctolero, 2024; Mantaring, 2024).

Apart from architecture, other physical characteristics of natural and man-made elements of the built environment influence the relative tourism value of a location. Maitland and Smith (2009) argue that these elements contribute aesthetically to a location's tourism potential. This is consistent with the argument of Specht (2014) where the desires and expected experiences of the visitors shape the development of a destination to be suitable for tourism. However, as in man-made structures, natural elements also share the problems of maintenance, care, and preservation. This is important also in the Philippine setting, as significant historical elements of the tourism portfolio are environmental in nature (Cagat, 2018; Cayron, 2016). These are also integral in the overall complementation of the tourism and built environment sectors.

Seaports

For regions such as North America and the Mediterranean Sea where cruise tourism is regarded a mature industry, expansion in ship sizes, services and entertainment onboard, new itineraries or ports of call in the passenger experience are given attention (Weeden et al., 2011), while those regarded part of emerging cruise regions endeavor to improve their infrastructures for international accessibility and diverse cruise passengers (Brida et al., 2012; Sun et al., 2014). This is in line with the ASEAN roadmap on cruise tourism, as well as the Philippine government's efforts in aligning local tourism programs through this initiative given the mobility of local tourists and those engaging in cruise travels with Southeast Asia and the rest of the world (Abueg, 2023a; Acejo & Abueg, 2022).

Under the PTDP 2016–2022, a dedicated cruise port and terminal in Metro Manila was established by 2018 to make way for fly-cruise programs and existing destination ports such as Manila, Boracay, Puerto Princesa and Coron (Palawan), Iloilo City, Tagbilaran City (Samar), Cebu City (Cebu), and Salomague Island (Ilocos Sur) – Currimao (Ilocos Norte) were also completed by 2020. Overall, it was indicated that the "infrastructure issues for cruise tourism, particularly in improving transit ports of call, are now being addressed but still exhibit limited potential until such time as a dedicated cruise terminal in Manila is put in place" (DOT, 2016, p. 6). The same strategy was reiterated through an improved transportation network through inter-sectoral linkages identified specifically in the seventh chapter of the PDP 2023–2028 (NEDA, 2023).

Landports

Land-based attractions within the periphery of ports of call also depend on the state of road infrastructure and networks significantly affecting the cruise experience. The experience of Bohol, one of the country's top tourist destinations, provides an insight on developing tourism capacity while ensuring the locals' concerns. As part of giving due consideration on the natural increase of its population along with an increase in tourist visitors, tourism planning in Bohol was essentially geared on providing the necessary infrastructure such as major roads (Acejo et al., 2004). Bureaucracies in infrastructure development, however, remain a topical concern as various attractions in Bohol province are still faced with "inconsistently connected local and national road networks" (Mirabueno & Yujuico, 2014, p. 313). Collaborations between government agencies according to the study are identified as crucial to having a road network that can accommodate the tourism vision. The Department of Public Works and Highways (DPWH's) Tourism Road Infrastructure Program has been tasked with connecting tourism sites, identifying routes and developing criteria for prioritizing farm tourism roads (DOT, 2016). The trajectory of development of tourism projects with agricultural landscapes can be however contentious when people and livelihoods are displaced such as in the case of Nasugbu, Batangas (Dizon, 2015).

Airports

The Asia–Pacific region contributed the highest share of global passenger traffic at 37% (Air Transport Action Group, 2020). Over the next two decades, air travel outlook shows a growing demand that hover at around 4.2% annually. Similar trend points to a rapidly expanding demand for air travel in the Philippines which currently has an aviation infrastructure score of 3.2 which is below the South East Asia average of 3.7 (1–7 score), from the World Economic Forum (WEF) Travel and Tourism Competitiveness Report (WEF, 2019). While the 2019 Travel and Tourism Competitiveness Index (TTCI) shows an increase in ranking to 59th compared to its 89th rank in 2009 in terms of the quality of air transport infrastructure, the Philippines remains in the bottom among Asia–Pacific countries.

Major capacity improvement in infrastructure relating to passenger comfort remains an issue. Airport terminal congestion, as the maximum passenger volume has been reached in Ninoy Aquino International Airport (NAIA) (principal international airport in the country), has been highlighted. The growing needs of the industry in infrastructure capacity along with issues on safety and ownership and control have been reported by the International Air Transport Association (IATA, 2015). An ongoing safety concern in operations such as air traffic coordination using technological advances and control procedure for efficient communication was raised as urgent. The inadequacy of the facilities is further affected by the domestic airports' practice of scheduling their flights during the day owing to a lack of equipment for night landing (JICA, 2011; in Francisco & Lim, 2022).

These destination infrastructures consequently require a contingency plan within which possible disasters and crisis events are mitigated. Current perspectives on resilience and crisis research have raised the minimal attention to planning and lack of sufficient resources, despite growing attention to its importance (Hall et al., 2023). Emergency preparedness in times of earthquakes can affect not only the physical infrastructure but also the "spatial, organizational characteristics and sectoral" domains of a community and its overall resilience (Hall & Prayag, 2021, p. 14). While the size and capacity of firms involved also demonstrate the level of response within which these concerns may be addressed (Ngin et al., 2020), coordination with the government (both regional and local) also helps in disaster recovery through, for example, identification of accountability of tourism stakeholders (e.g., Orchiston & Higham, 2016).

Travel and Tourist Experience

As a complement of the tourism experience given the elements of the built environment, travel is also a recognized component of this nexus. Ewing and Cervero (2010) looked at the importance of modes of transportation to complement and enhance the tourism experience of visitors by surveying literature through a meta-analysis. This recognition is important since tourists' experience in travel from one destination to next (and from and to their home destination) affects the relative tourism value of the destination. In terms of the nonphysical elements of the built environment, as we have seen through the course of the COVID-19 pandemic, digitization has been hastened and accelerated in almost all facets of economic and social spheres of society across the world and benefitted tourism and created nonmonetary models and transaction platforms (Abueg, 2020a, 2021, 2023b).

"Bridges" of Development From Tourism and Built Environment Sectors to Local Economies

In the age of postpandemic digital world, emphasis on reliable information is the backdrop of sustainable, resilient, and future-proof programs and policies for the economy, business, and other institutions of society. In the initial years of the pandemic, there were discussions on whether to use and mandate vaccine passports for travelers. This would greatly affect the opening of the tourism industry – from local tourism to both intra-ASEAN and extra-ASEAN visitors, as well as interoperability of vaccines used across the globe (Greene et al., 2021; Markovitz, 2021). Given that there is a platform of the "single-visa" and "open skies" brought forth before the pandemic, the additional health protocols became a game changer in further promoting this plan for the region. Further, there was an uneven access to initially produced vaccines with emergency-use authorization, with the Philippines to be the last to implement in the ASEAN (Amarila & Abueg, 2022). The absence – and later uneven access – of vaccines has indicated significant levels of uncertainty across economic sectors and affects the path to full reopening of the economy while not sacrificing the overall health of the citizens (Abueg, 2020b; Pernia, 2020).

Together with the harmonization of policies postpandemic, existing physical infrastructure must be revisited and retrofitted to be compliant with health protocols and practices (APEC, 2021; ASEAN, 2021). The core of the built environment must also be in unison with the postpandemic norms and practices and to ensure that the gradual recovery and redefined sustainability of tourism will complement the existing elements of the built environment, such as infrastructure, cross-border mobility, and travel (Ewing & Cervero, 2001, 2010; Pellegrino, 2021). For example, the identified necessary infrastructure requirements for promoting cruise tourism and local tourism and travel must be revisited and improved to ensure tourism demand and later economic value added (Holland et al., 2021; Zhang et al., 2022). Further, the reopening of the economy due to increase and surge of travels has led to what is currently coined as "revenge travels" – which must be aided with adequate, correct, and health-compliant infrastructure. Not only the infrastructure for tourism and travel must be reviewed but also the existing infrastructure in destinations, the reception and capability of the tourism workers postpandemic, and ensuring the implementation of new policies and protocols. In the Philippines alone, there is an indication of a bounce back of cruise tourism, which is also seen to be in unison with the ASEAN's plan of making the region a single destination for cruise (Philippine News Agency, 2021; Rocamora, 2023).

To align the new protocols, generation of larger data sets in aid of sound policy and decision-making plus the revisit of the elements of the built environment became the standard requisites. Further, digital technology and infrastructure became the name of the game as the world navigates itself out of the pandemic. This is also equally important in tourism, and we see that businesses that had the capacity to complement their prepandemic operations with digital platforms have a higher chance of business survival and have potential for innovation. Because of the pandemic, a lot of enterprises – mostly microenterprises – have been affected by the halt of human mobility and economic activities not only in the Philippines (Abueg, 2020b, 2021) but also in the ASEAN and the world (UNESCO, 2021). Digitization has shown a lot of potential in the revival and recovery of tourism sector, especially in ASEAN (UNESCO, 2021), which is consistent with the recognized situation and roadmap of the ten-member bloc (ASEAN Secretariat, 2021a).

Survey of Local Tourism Sectors and Built Environment Complementation

In the complementation of the tourism and built environment sectors – from the local to the international arena – we see common themes that are necessary for future-proofing the tourism and built environment sectors and ensuring their seamless complementation. In this section, we indicate some of the possible leads for policies and steps that may be taken for further growth, resilience, and sustainability. Apart from the ministerial statement of the ASEAN in 2021 (ASEAN Secretariat, 2021a), a more comprehensive and larger in scope plan and declaration was made in the same year by the East Asia Summit (ASEAN Secretariat, 2021b) that specifically outlines the identified strategies for the recovery of the tourism sector.

The most prominent outcome early in the pandemic is the impact on seafarers' employment situation where seafarers are frequently reported being stranded ashore and aboard. Global processes at sea such as an outsourced workforce under temporary employment contract have been a pervading feature aboard cargo and cruise ships that presents sea-based workers with precarious employment conditions (Aguilar, 2014; Chin, 2008; DeSombre, 2006). As a major source of sea-based workers aboard ships, the Philippine's role in providing for a fair and decent work, not least sustainable income source (Acejo & Abueg, 2022), is a key to understanding the vaunted claims of opportunities in cruise tourism. As the government's stance between maintaining a niche in the seafaring global labor market and upholding the work conditions of seafarers has not been straightforward, the onslaught of the pandemic has exposed how this conflicting position can be a source of vulnerability particularly in the case of its cruise workers who were faced with job losses. The lack of bargaining power within which Filipino cruise workers could have been protected has been undermined by a standard employment contract (Terry, 2009), often temporary, which makes it difficult to push for decent and fair labor standards particularly for the Philippines where labor regulations offer limited protection for its workers (Dacanay & Walters, 2011). In instances where employment relationship is relatively more established such as the case for British seafarers having permanent contract, the problematic role of the shipping companies described as "too willing to negatively alter the employment terms and conditions" onboard during the pandemic was regarded too as less than ideal (Devereux & Wadsworth, 2021, p. 16).

Unsurprisingly, cruise workers' overall health and safety challenges spurred attention among maritime scholars. The increased number of suicides during the early stages of the pandemic has highlighted the mental health challenges and other occupation-related issues of cruise workers (see, for example, Radic et al., 2020). The body of work however has little attention on the extent to which feminization of working practices aboard cruise ships has safety and well-being implications in terms of tensions brought about by sexual harassment, abuse, and sexually transmitted infections (Belcher et al., 2001; Thomas, 2003).

Experiences of countries with a mature cruise tourism industry point to financial justification of the nature of port infrastructure that will accommodate cruise ships. Destination ports regarded as a port of call (and not a home port) may prove to be more costly as gains from cruise ship presence may not always be quite lucrative (Vayá et al., 2018). Along with this, the environmental and social impacts on the local, if not remote, communities also require careful consideration as these also set the tone on the extent to which cruise tourism may be promoted and encouraged by the locals themselves. Postpandemic lessons arising from Florida's heavy reliance on cruise tourism also provide guidance on the importance of taking stock of all stakeholders – destination planners and their municipalities, cruise industry executives, salesforce, companies in the supply chain, passengers, and crew members (da Silva, 2021), as a responsible way forward that can create a sustainable approach to cope with the predicted rapid demand of cruise tourism.

Apart from the importance of tourism workers, there must be also recognition of potential workers that may be deployed into boosting tourism and built environment complementation. In the Philippines, Correa and Abueg (2020) suggested that the program of the government to bring back displaced workers in the provinces must be utilized to the advantage of the built environment sector outside the National Capital Region. It is also crucial in this reintegration efforts to profile labor losses due to the pandemic (Epetia, 2020), so that we can place them correctly in jobs that they can maximize their productivity and potentials.

The elements of built environment that we see are not limited to man-made infrastructure of physical structures but also include the natural elements that come with the destination. Given these, one facet of the nexus of tourism and built environment is to enhance these environmental elements that will continue to affirm experience and increase tourism potential and value, as mentioned earlier in Maitland and Smith (2009) and Specht (2014). For example, the world-renowned Guimaras mangoes are the first to receive geographical indication (GI) issued by the Intellectual Property of the Philippines or IPOPHL.[3] GIs are aligned and consistent with the 2015 Geneva Act of the Lisbon Agreement (Talavera, 2023). In the same year, the island of Bohol became the first global geopark of the UNESCO. This highlights the natural treasure of the tourist destination of the island, and the island enlists itself with other 194

[3]In 1995, the Guimaras mango is awarded the "sweetest fruit" by the Guinness World Records.

geoparks in 48 countries (Adel, 2023b). Notably, the island province of Bohol has been capitalizing on its tourism portfolio, which had been a notable destination for tourism development research (Acejo et al., 2004).

The accelerated digitization is one major theme of the roadmaps for economic recovery across the world. As mentioned earlier, this digitization complementing existing businesses created more productive and resilient business models, with potentials for innovation and future growth. In tourism, we see this even before the pandemic (UNWTO, 2018) and is seen to continue even during the post-pandemic period (UNWTO, 2021). Such recognition of this postpandemic built environment ingredient must be planned carefully and strategically, especially complementing the businesses of the tourism sector (Galán et al., 2022). We also highlight that policymakers must also provide support especially for micro-enterprises in terms of procuring these new implements and ensuring that the benefits are enjoyed by most of the sector and the economy (OECD, 2021).

One consequence of the pandemic is the change in travels: from long to short travels, plus the use of more electronic-enabled vehicles that support the environmental healing because of the sudden stops in human activities. In the Philippines and in many parts of the world, one mode of transportation that intensified as we navigate postpandemic is the use of bicycles (People Powered Movement, 2020). For example, the Philippines managed to quickly adjust major arterial conduits of transportation to incorporate bicycle lanes (Xu & Mayuga, 2022). This intensified use of bicycles to address shortage in transportation and increasing costs of gasoline even created another form of tourism complementary to the built environment. The now-called "bicycle tourism," which includes bicycles as mode of transportation and travels for tourists and reaching tourist destinations, has also defined places which are creating places for such travel (Adventure Cycling Association, 2023). While this is a subset of active transport tourism, Specht (2014) emphasizes that travel and transport are integral elements in tourism, and may further enhance tourism and natural environment portfolios.

Another form of transportation that became intensely used to complement tourism and related travels is the use of electronic tricycles (or "e-trikes"). Although generally limited in terms of ability in mileage, e-trikes became a transitory feature in a lot of tourist destinations. It mitigates not only carbon emissions of traditional (or gasoline-powered tricycles) but also reduces noise pollution, both for the traveler and for the destination. This iteration on tourism travel has been promoted even before the pandemic and is continued to be of use and adoption in several destinations (Subingsubing, 2019). Such is aligned to the national government's mandate of measuring sustainability in the tourism sector. The national government's Measuring Sustainability for Tourism (MST) tracks water consumption, energy consumption, and carbon emissions from energy use. With the recovering of the tourism sector, the MST indicators all have shown increasing trends (Fig. 9.2).

In all, these developments are replete with recalibration of existing models in practices, within the tourism and built environment sectors, as well as looking at the interactions of these sectors. There is a recognition that not only the economic, social, and environmental aspects of the sectors are important but also cultural, heritage, and stakeholders' inputs that transform tourism and its complementation with the

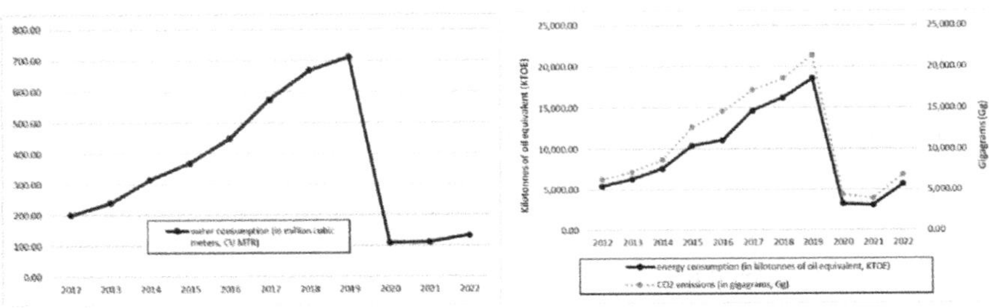

Fig. 9.2. [Left] Total Water Consumption of Tourism Sector (Inbound and Domestic) in Million Cubic Meters, 2012–2022. [Right] Total Energy Consumption of Tourism Sector (Electricity, Petroleum, and Other Oil Products) in Kilotonnes of Oil Equivalent (KTOE) and Carbon Dioxide (CO_2) in Gigagrams, 2012–2022. *Source:* PSA.

built environment (Acri, 2022; Chuang, 2023; Han et al., 2023; Lazar & Chithra, 2022; Ngo & Creutz, 2022; Rus et al., 2022; Xu et al., 2022). It is also important to emphasize that we must understand the particularities of a local tourism portfolio and the influences on built environment, in ensuring complementation with its elements. In this way, proper interventions through policies, programs, laws, and even stakeholder initiatives may be implemented to enhance renewed sustainability, resilience, and future-proofing given the backdrop of the pandemic experience.

Afterword: Future-Proofing Tourism and Built Environment Nexus Toward Postpandemic Sustainability

The COVID-19 pandemic has affected significantly the tourism and built environment sectors, and in the Philippines where these sectors are economically relied on for support, these have crucial implications. With the "next normal" in the offing, technology and digitization has become the game changer that has modified the models of business, economy, governance, and society. The pandemic also necessitated a revisit of plans, policies, and programs for sustainability, resiliency, and future-proofing of tourism and its nexus with other sectors – the built environment sector.

In addition to these changes, we recognize that moving forward, the nexus of tourism and built environment sectors is not working within a vacuum. The importance of multi-stakeholder participation and improved governance across the bureaucracy must also be a crucial consideration. Farsari (2021) noted that these approaches to tourism, such as the complexity adaptive systems and complexity theory, are increasingly referred to in recent studies (Dredge, 2019; Farrell & Twining-Ward, 2004; Lélé, 1998; in Farsari, 2021). These may also be applied in the built environment sector, as the pandemic experience in the Philippines also indicated such elements in the built environment sector (Abueg et al., 2021, 2022;

Correa & Abueg, 2020). We also highlight the importance of appropriate recalibration and customization of various types of tourism and built environment complementarities across destinations and portfolios to further resilience and sustainability.

This recognition and the Philippine experience are in unison with the arguments of Farsari (2021) on having an evolving tourism governance while considering the increasing complexity of the sectors in question and the research that one might undertake. The road to resiliency, albeit identified in the United Nations Sustainable Development Goals, has little indications about how it would be pursued (Sharpley, 2020; in Farsari, 2021). This is important given that the Philippines is in unison with the programs of ASEAN for tourism across the region. Finally, these indications and experiences should be taken into consideration and account in planning and policymaking in terms of the potential advantages and silver linings despite the negative effects and externalities brought forth by the pandemic.

References

Abueg, L. C. (2020a). Silver linings in Philippine history and macroeconomics of the Covid-19 pandemic response: Beyond the longest lockdown. *Philippine Journal of Health Research and Development*, *24*(4), 50–61.

Abueg, L. C. (2020b). Extended, enhanced, and extreme: Macroeconomic implications of the community quarantine in the Philippines due to the COVID-19 pandemic. In *Economics and management matters no. 2020-01*. University of the Philippines Los Baños.

Abueg, L. C. (2021). What can gender economics learn from the Pinoy BL genre of the Covid-19 pandemic? *Review of Women's Studies*, *31*(1), 35–62.

Abueg, L. C. (2023a). ASEAN integration and tourism: Towards a sustainable and inclusive regional tourism program. In J. P. R. Rivera, M. C. L. S. Rodolfo, F. M. Y. Roxas, & E. L. M. Gutierrez (Eds.), *People, planet, profit: Principles and practices of sustainable tourism* (pp. 431–464). Joint publication of the Asian Institute of Management (Makati City), & De La Salle University Publishing House (City of Manila).

Abueg, L. C. (2023b). Stretching Filipino "resilience" amidst a pandemic-induced recession with market and government failures: Insights from community pantries and online barter. *Graduate Inequality Review*, *2*(1), 68–76.

Abueg, L. C., Amarila, M. R. M., Correa, L. N. V., & Zamora, C. M. B. (2022). REIT we trust? Revisiting the trust fund towards the "new normal" of the Philippine built environment sector. In *Proceedings of the 15th national convention on statistics*. https://psa.gov.ph/ncs/paper-presentations-and-manuscripts/15th

Abueg, L. C., Zamora, C. M. B., & Correa, L. N. V. (2021). Covid-19 pandemic and the Philippine real estate property cycle: Indications of bubble and burst? *Philippine Review of Economics*, *58*(1&2), 265–293.

Acejo, I. L., & Abueg, L. C. (2022). Sailing through post-pandemic recovery? The role of the state for workers in cruise tourism industry in the Philippines. In P. Mohanty, A. Sharma, J. Kennell, & A. Hassan (Eds.), *The Emerald handbook of destination recovery in tourism and hospitality* (pp. 227–248). Emerald Publishing Limited.

Acejo, I. L., del Prado, F., & Remolino, D. (2004). Tourism fuels an emerging city: The case of Tagbilaran City, Bohol. In *PIDS discussion paper series no. 2004-53*. https://pids.gov.ph/publications/category/discussion-papers

Acri, M. (2022). Designing conservation: From sustainable heritage to sustainable tourism (and vice versa). In A. Mandić, R. A. Castanho, & U. Stankov (Eds.), *Cultural sustainable tourism. Advances in science, technology & innovation*. Springer. https://doi.org/10.1007/978-3-031-10800-6_13

Adel, R. (2023a, May 22). 'Built heritage emergency': Advocates lament fire at Manila central post office. *Philippine Star*. https://www.philstar.com/headlines/2023/05/22/2268166/built-heritage-emergency-advocates-lament-fire-manila-central-post-office

Adel, R. (2023b, May 25). Bohol named the Philippines' first UNESCO global geopark. *Philippine Star*. https://www.philstar.com/lifestyle/travel-and-tourism/2023/05/25/2268942/bohol-named-philippines-first-unesco-global-geopark

Adventure Cycling Association. (2023). *Bicycle tourism 101*. https://www.adventurecycling.org/advocacy/building-bike-tourism/bicycle-tourism-101/

Agcaoili, L. (2023, May 20). Too early to talk about rate cuts. *Philippine Star*. https://www.philstar.com/business/2023/05/20/2267537/too-early-talk-about-rate-cuts

Aguilar, F. V. (2014). *Migration revolution: Philippine nationhood and class relations in a globalized age*. Ateneo de Manila University Press.

Air Transport Action Group. (2020). *Aviation: Benefits beyond borders*. ATAG.

Alcazaren, P. (2024, January 20). A night at the Pasig. *Philstar Life*. https://philstarlife.com/geeky/584044-pasig-river-esplanade

Amarila, M. R. M., & Abueg, L. C. (2022). Do Philippine stocks catch coronavirus? Some econometric check-up on pandemic data, 2021–2022. *Journal of Economics, Management and Agricultural Development*, *8*(1), 55–70.

APEC. (2021). *APEC cross-border travel: Roundtable on safe passage, 7 May 2021 – Report to Senior Officials*. http://mddb.apec.org/Documents/2021/SOM/SOM3/21_som3_007.pdf. Accessed on January 16, 2024.

ASEAN. (2021, May 7). *Final study report for the post-Covid-19 recovery plan for ASEAN tourism*. https://asean.org/wp-content/uploads/2021/09/Final-Study-Report_Post-Covid19-Tourism-Recovery-Plan-for-ASEAN-Tourism_....pdf

ASEAN Secretariat. (2018). *Twenty-first meeting of ASEAN tourism ministers joint media statement*. https://asean.org/wp-content/uploads/2018/01/JMS-M-ATM-21-Final.pdf

ASEAN Secretariat. (2021a, January 25). *Joint media statement of the ASEAN Tourism Ministers on the post Covid-19 recovery plan for ASEAN tourism*. https://asean.org/wp-content/uploads/2021/09/JMS-Post-COVID-19-Recovery-Plan-for-ASEAN-Tourism-Final-00000002.pdf

ASEAN Secretariat. (2021b, October 27). *East Asia summit leaders' statement on economic growth through tourism recovery*. https://asean.org/wp-content/uploads/2021/10/99.-EAS-Leaders-Statement-on-Economic-Growth-through-Tourism-Recovery-FINAL.pdf

Belcher, P., Lane, A. D., Sampson, H., Thomas, M., Veiga, J., & Zhao, M. (2001). *Women seafarers: Global employment policies and practices*. Seafarers International Research Centre report for the ILO.

Boiser, A., Mangosing, F., & Aurelio, J. M. (2023, May 23). Calls for rebuilding central post office ring out after fire. *Philippine Daily Inquirer*. https://newsinfo.inquirer.net/1772871/calls-for-rebuilding-central-post-office-ring-out-after-fire

Breene, K. (2016, May 26). 7 surprising things you probably don't know about ASEAN. *World Economic Forum*. https://www.weforum.org/agenda/2016/05/7-surprising-things-about-asean

Brida, J. G., Pulina, M., Riaño, E., & Zapata-Aguirre, S. (2012). Cruise passengers' experience embarking in a Caribbean home port: the case study of Cartagena de Indias. *Ocean & Coastal Management*, *55*, 135–145.

Cabalza, D. (2020, February 18). PH got record 8 million tourists in 2019. *Philippine Daily Inquirer*. https://newsinfo.inquirer.net/1230026/ph-got-record-8-million-tourists-in-2019

Cagat, K. A. D. (2018). Mixed views on the Philippines' Ifugao Rice Terraces: 'Good' versus 'beautiful' in the management of a UNESCO world heritage site. *Journal of Southeast Asian Studies*, *49*(1), 84–104. https://doi.org/10.1017/S0022463417000704

Capistrano, R. C., & Muñoz, K. E. (2022). Tourism mobilities in the Philippines: A historiographical analysis of travel and tourism activities from pre-1950s. *Journal of Tourism History*, *14*(1), 28–46. https://doi.org/10.1080/1755182X.2022.2085817

Caynila, K. A. M., Luna, K. T., & Milla, A. A. (2022). The Philippine tourism sector amid the pandemic: Developments and prospects. *Bangko Sentral ng Pilipinas Economic Newsletter no. 22-02*. https://www.bsp.gov.ph/Media_And_Research/Publications/EN22-02.pdf

Cayron, J. G. (2016). Archaeological heritage tourism in the Philippines: Challenges and prospects. In P. Gould & K. Pyburn (Eds.), *Collision or collaboration. One world archaeology*. Springer. https://doi.org/10.1007/978-3-319-44515-1_7

Cervantes, F. M. (2021, June 4). 'One town, one product' bill gets final house nod. *Philippine News Agency*. https://www.pna.gov.ph/articles/1142619

Chin, C. B. N. (2008). Labour flexibilization at sea: 'Mini U[nited] N[ations]' crew on cruise ships. *International Feminist Journal of Politics*, *10*(1), 1–18.

Chuang, C. M. (2023). The conceptualization of smart tourism service platforms on tourist value co-creation behaviours: An integrative perspective of smart tourism services. *Humanities & Social Sciences Communications*, *10*(367). https://doi.org/10.1057/s41599-023-01867-9

Correa, L. N. V., & Abueg, L. C. (2020). The Philippines' Balik Probinsya, Bagong Pag-asa initiative and the real estate development sector during the COVID-19 pandemic: Some policy morals. *International Journal of Reconfigurable and Embedded Systems*, *14*(S2), 156–176.

da Silva, A. L. R. (2021). An overview of the impact of COVID-19 on the cruise industry with considerations for Florida. *Transportation Research Interdisciplinary Perspectives*, *10*, 100391.

Dacanay, J., & Walters, D. (2011). Protecting precarious workers in the global maritime industry: A case of regulatory failure? *Policy and Practice in Health and Safety*, *9*(2), 47–68. https://doi.org/10.1080/14774003.2011.11667761

DeSombre, E. R. (2006). *Flagging standards. Globalization and environmental, safety, and labor regulations at sea*. MIT Press.

Devereux, H., & Wadsworth, E. (2021). *The forgotten keyworkers: Challenges faced by British seafarers as a result of the Covid-19 pandemic*. Solent University. https://www.solent.ac.uk/research-innovation-enterprise/documents/forgotten-keyworkers-project-report-september-2021.pdf

Dizon, H. M. (2015). The contested development of a Philippine tourism landscape: The case of Nasugbu, Batangas. *Kasarinlan: Philippine Journal of Third World Studies*, *30*(1), 91–129.

Doctolero, J. (2024, January 18). Pasig river transformed: Filipinos to reap benefits from PBBM's ambitious urban renewal. *Philippine Information Agency*. https://pia.gov.ph/news/2024/01/18/pasig-river-transformed-filipinos-to-reap-benefits-from-pbbms-ambitious-urban-renewal

DOT. (2016). *National tourism development plan 2016–2022*. Department of Tourism. https://visitcentralluzon.com/wp-content/uploads/2018/11/NTDP-2016-2022_Executive-Summary.pdf

Dredge, D. (2019). Governance, tourism, and resilience: A long way to go? In J. Saarinen & A. M. Gill (Eds.), *Resilient destinations and tourism: Governance strategies in the transition towards sustainability in tourism* (pp. 48–66). Routledge.

Epetia, M. C. F. (2020). COVID-19, job loss, and underemployment: Who is affected? *Philippine Review of Economics*, *58*(1&2), 63–91.

Ewing, R., & Cervero, R. (2001). Travel and the built environment: A synthesis. *Transportation Research Record: Journal of the Transportation Research Board*, *1780*(1), 87–114. https://doi.org/10.3141/1780-10

Ewing, R., & Cervero, R. (2010). Travel and the built environment: A meta-analysis. *Journal of the American Planning Association*, *76*(3), 265–294. https://doi.org/10.1080/01944361003766766

Farrell, B. H., & Twining-Ward, L. (2004). Reconceptualising tourism. *Annals of Tourism Research*, *31*(2), 274–295. https://doi.org/10.1016/j.annals.2003.12.002

Farsari, I. (2021). Exploring the nexus between sustainable tourism governance, resilience and complexity research. *Tourism Recreation Research*, *48*(3), 352–367. https://doi.org/10.1080/02508281.2021.1922828

Francisco, K. A., & Lim, V. L. (2022). *Philippine air transport infrastructure: State, issues, government strategies*. Philippine Institute for Development Studies Discussion. Paper no. 2022-62. https://pids.gov.ph/publications/category/discussion-papers

Galán, J. J., Carrasco, R. A., & Latorre, A. (2022). Strategic digitization of tourism. In J. L. Reis, E. P. López, L. Moutinho, & J. P. M. D. Santos (Eds.), *Marketing and smart technologies. Smart innovation, systems and technologies* (Vol. 279). Springer. https://doi.org/10.1007/978-981-16-9268-0_43

Greene, D. N., McClintock, D. S., & Durant, T. J. S. (2021). Interoperability: COVID-19 as an impetus for change. *Clinical Chemistry*, *67*(4), 592–595. https://doi.org/10.1093/clinchem/hvab006

Hall, C. M., & Prayag, G. (2021). Earthquakes and tourism: Impacts, responses and resilience – An introduction. In C. M. Hall & G. Prayag (Eds.), *Tourism and earthquakes* (pp. 1–35). Channel View Publications.

Hall, C. M., Safonov, A., & Naderi Koupaei, S. (2023). Resilience in hospitality and tourism: Issues, synthesis and agenda. *International Journal of Contemporary Hospitality Management*, *35*(1), 347–368.

Han, C., Zhou, L., & Zhou, T. (2023). How does the built environment affect hotel prices? A study using multiscale GWR and deep learning. *Journal of Asian Architecture and Building Engineering*. https://doi.org/10.1080/13467581.2023.2270027

Holland, J., Mazzarol, T., Soutar, G. N., Tapsall, S., & Elliott, W. A. (2021). Cruising through a pandemic: The impact of COVID-19 on intentions to cruise.

Transportation Research Interdisciplinary Perspectives, *9*, 100328. https://doi.org/10.1016/j.trip.2021.100328

IATA. (2015, May). *Immediate and long-term priorities for Manila airports, Manila, Philippines*. https://documents.pub/document/immediate-and-long-term-priorities-for-manila-a-contentassets-a-ee788aa527d444dfad.html?page=1

Ines, J. (2023, May 23). Unfortunate events besetting Manila's heritage buildings in recent years. *Rappler*. https://www.rappler.com/newsbreak/iq/events-besetting-manila-heritage-buildings/

JICA. (2011). *Study on the airport strategy for the greater capital region in the republic of the Philippines*. Japan International Cooperation Agency.

Lazar, N., & Chithra, K. (2022). Role of culture in sustainable development and sustainable built environment: A review. *Environment, Development and Sustainability*, *24*, 5991–6031. https://doi.org/10.1007/s10668-021-01691-8

Lélé, S. (1998). Resilience, sustainability, and environmentalism. *Environment and Development Economics*, *3*(2), 249–254. https://doi.org/10.1017/S1355770X98260128

Licsi, A. (2024, January 3). Razed Manila post office to become a tourist destination after its restored. *Philstar Life*. https://philstarlife.com/news-and-views/508967-razed-manila-post-office-to-become-tourist-destination

Maglinong, D. (2018, May 28). What we could lose in the fire that hit the National archives. *Interaksyon*. https://interaksyon.philstar.com/breaking-news/2018/05/28/127600/what-we-could-lose-in-the-fire-that-hit-the-national-archives/

Magramo, K. (2023, May 21). Inferno tears through Manila's historic central post office. *CNN*. https://edition.cnn.com/2023/05/21/asia/fire-manila-central-post-office-intl-hnk/index.html

Maitland, R., & Smith, A. (2009). Tourism and the aesthetics of the built environment. In J. Tribe (Ed.), *Philosophical issues in tourism*. Channel View Publications.

Mantaring, J. R. (2024, January 17). Marcos vows 'people-centered development' in his Pasig River restoration project. *CNN Philippines*. https://www.cnnphilippines.com/news/2024/1/17/marcos-pasig-river-restoration-project.html

Markovitz, G. (2021, May 5). What is a 'vaccine passport' and will you need one the next time you travel? *World Economic Forum*. https://www.weforum.org/agenda/2021/05/what-is-a-vaccine-passport-and-will-you-need-one-the-next-time-you-travel/

Mirabueno, J., & Yujuico, E. (2014). Paving the way for Philippine tourism via interagency collaboration on road networks. *Transport Policy*, *36*, 306–315.

NEDA. (2023). *Philippine development plan 2023–2028*. https://pdp.neda.gov.ph/philippine-development-plan-2023-2028/

Ngin, C., Chhom, C., & Neef, A. (2020). Climate change impacts and disaster resilience among micro businesses in the tourism and hospitality sector: The case of Kratie, Cambodia. *Environmental Research*, *186*, 109557. https://doi.org/10.1016/j.envres.2020.109557

Ngo, T. H., & Creutz, S. (2022). Assessing the sustainability of community-based tourism: A case study in rural areas of Hoi An, Vietnam. *Cogent Social Sciences*, *8*(1), 2116812. https://doi.org/10.1080/23311886.2022.2116812

OECD. (2021). Preparing tourism businesses for the digital future. In *OECD tourism trends and policies 2020*. OECD Publishing. https://doi.org/10.1787/6b47b985-en

Ong, G. (2022, June 18). Fire hits iconic metropolitan theater in Manila. *Philippine Star*. https://www.philstar.com/nation/2022/06/18/2189207/fire-hits-historic-metropolitan-theater

Orchiston, C., & Higham, J. E. S. (2016). Knowledge management and tourism recovery (de) marketing: the Christchurch earthquakes 2010–2011. *Current Issues in Tourism*, *19*(1), 64–84.

Pellegrino, F. (2021). Tourism, transport and sustainability. In F. Grasso & B. S. Sergi (Eds.), *Tourism in the Mediterranean Sea* (pp. 279–289). Emerald Publishing Limited. https://doi.org/10.1108/978-1-80043-900-920211019

People Powered Movement. (2020). *Cities around the world are adding bike lanes thanks to COVID-19*. https://www.peoplepoweredmovement.org/cities-around-the-world-are-adding-bike-lanes-thanks-to-covid-19/

Pernia, E. M. (2020). *Health is economy: Some lessons from COVID-19*. University of the Philippines School of Economics Discussion Paper no. 2020-13. https://econ.upd.edu.ph/dp/index.php/dp/article/view/1536/1020

Philippine News Agency. (2021, February 5). DOT welcomes Asean as single tourism destination. https://www.pna.gov.ph/articles/1129728

Pilapil-Añasco, C., & Lizada, J. C. (2014). Philippine tourism: Evolution towards sustainability. *SHS Web of Conferences*, *12*, 01032. https://doi.org/10.1051/shsconf/20141201032

Radic, A., Lück, M., Ariza-Montes, A., & Han, H. (2020). Fear and trembling of cruise ship employees: Psychological effects of the COVID-19 pandemic. *International Journal of Environmental Research and Public Health*, *17*(18), 6741.

Rocamora, J. A. L. (2023, February 15). Cruise tourism back after 3 years; 139 port calls in PH expected. *Philippine News Agency*. https://www.pna.gov.ph/articles/1195304

Rus, K. A., Dezsi, S., Ciascai, O. R., & Pop, F. (2022). Calibrating evolution of transformative tourism: A bibliometric analysis. *Sustainability*, *14*, 11027. https://doi.org/10.3390/su141711027

Sarao, Z. (2021, September 5). Fire hits part of PH National archives building; archival collections safe. *Philippine Daily Inquirer*. https://newsinfo.inquirer.net/1483703/fire-hits-part-of-ph-national-archives-building-archival-collections-safe

Sharpley, R. (2020). Tourism, sustainable development and the theoretical divide: 20 years on. *Journal of Sustainable Tourism*, *28*(11), 1932–1946. https://doi.org/10.1080/09669582.2020.1779732

Specht, J. (2014). Tourism and the built environment. In *Architectural tourism*. Springer Gabler. https://doi.org/10.1007/978-3-658-06024-4_2

Subingsubing, K. (2019, September 13). LTFRB allows e-trikes as tourist transport. *Philippine Daily Inquirer*. https://newsinfo.inquirer.net/1164095/ltfrb-allows-e-trikes-as-tourist-transport

Sun, X., Feng, X., & Gauri, D. K. (2014). The cruise industry in China: Efforts, progress and challenges. *International Journal of Hospitality Management*, *42*, 71–84.

Talavera, C. (2023, June 3). 'Sweetest' Guimaras mango becomes first geographical indication in Philippines. *Philippine Star*. https://www.philstar.com/business/2023/06/03/2271012/sweetest-guimaras-mango-becomes-first-geographical-indication-philippines

Terry, W. C. (2009). Working on the water: On legal space and seafarer protection in the cruise industry. *Economic Geography*, *85*(4), 463–482.

Thomas, M. A. (2003). Sexual health of women working aboard cruise ships. *Health Education Journal, 62*(4), 293–303. https://doi.org/10.1177/001789690306200402

UNESCO. (2021, April 26). The future of tourism and culture of peace in ASEAN+ (with update in 2023). https://www.unesco.org/en/articles/future-tourism-and-culture-peace-asean

UNESCO. (2023, March 1). Cutting edge: Bringing cultural tourism back in the game. https://www.unesco.org/en/articles/cutting-edge-bringing-cultural-tourism-back-game

UNSD. (2016). *Measuring sustainable tourism: Developing a statistical framework for sustainable tourism*. UN World Tourism Organization. https://unstats.un.org/unsd/envaccounting/londongroup/meeting22/MST%20Overview%20of%20the%20initiative%2028Jul2016.pdf

UNSD. (2022). Measuring the sustainability of tourism (MST): Current state of play. In *UN world tourism organization report for the 53rd session of the UN statistical commission*. https://unstats.un.org/unsd/statcom/53rd-session/documents/BG-3l-MST-E.pdf

UNWTO. (2018, September). *Thematic discussion: 'Tourism and the digital transformation'*. https://www.e-unwto.org/doi/abs/10.18111/unwtoecd.2018.1.g51w645001604520

UNWTO. (2021, July 9). *Digital tools to revitalize tourism*. https://www.unwto.org/news/digital-tools-to-revitalize-tourism

Vayá, E., Garcia, J. R., Murillo, J., Romaní, J., & Suriñach, J. (2018). Economic impact of cruise activity: The case of Barcelona. *Journal of Travel & Tourism Marketing, 35*(4), 479–492.

Weeden, C., Lester, J. A., & Thyne, M. (2011). Cruise tourism: Emerging issues and implications for a maturing industry. *Journal of Hospitality and Tourism Management, 18*(1), 26–29.

WEF. (2019). *World economic forum travel and tourism competitiveness report 2019*. https://www.weforum.org/publications/the-travel-tourism-competitiveness-report-2019/

Xu, F., & Mayuga, K. (2022). From crisis to opportunity: How the Philippines built 500km of bike lanes in less than a year. *World Bank Blogs*. https://blogs.worldbank.org/transport/crisis-opportunity-how-philippines-built-500km-bike-lanes-less-year

Xu, G., Zhong, L., Wu, F., Zhang, Y., & Zhang, Z. (2022). Impacts of micro-scale built environment features on tourists' walking behaviors in historic streets: Insights from Wudaoying Hutong, China. *Buildings, 12*, 2248. https://doi.org/10.3390/buildings12122248

Zhang, H., Wang, Q., Chen, J., Rangel-Buitrago, N., & Shu, Y. (2022). Cruise tourism in the context of COVID-19: Dilemmas and solutions. *Ocean & Coastal Management, 228*, 106321. https://doi.org/10.1016/j.ocecoaman.2022.106321

Chapter 10

Pedaling Toward a Sustainable Intramuros: Assessing Cycle Tourism's Impact on Cultural Awareness and Environmental Preservation

Lilibeth C. Aragon, Jocelyn Y. Camalig and Ivy Charisse T. Pagulayan

Lyceum of the Philippines University Manila, Philippines

Abstract

This chapter focuses on cycle tourism using Bambike (Bamboo Bike) as sustainable transportation in the heritage site of Intramuros to promote cultural awareness and environmental preservation. Bambike is a bicycle where parts are made from natural bamboo and abaca with aircraft grade alloys tested for durability that passed the standards in the industry. The purpose of this chapter is to determine the value of cycle tourism using Bambike to promote green tourism in Intramuros and preserve its attractions. Its significance and benefits as to economic, environmental, and sociocultural impacts have been analyzed. The information source and data collection for this research was from the selected local and international tourists in Intramuros who have experienced and booked a trip with Bambike. An interview with structure questionnaire was done for the tourists and Bambike staff. It was concluded that Bambike can contribute to create site awareness leading to desire to preserve the heritage sites of Intramuros.

Keywords: Cycle tourism; Bambike; cultural awareness; environmental preservation; heritage sites

Revisiting Sustainable Tourism in the Philippines, 181–196

doi:10.1108/978-1-83753-678-820241010

Introduction

The current tourism landscape is undergoing a seismic shift, as new breed of travelers throws off the shackles of passive sightseeing and demands experiences that resonate on a deeper level. These intrepid souls, fueled by a yearning for self-discovery and meaningful connection, are no longer content with fleeting glimpses from behind bus windows. They crave immersion, challenge, and transformation, experiences that weave themselves into the very fabric of their being. This burgeoning desire propels the rise of cycling tourism, an activity that transcends mere transportation and thrusts riders into the heart of landscapes and cultures, forging memories not in snapshots but in the sweat, grit, and exhilaration of the journey itself.

Transportation is an integral part of the tourism industry, an integral thread woven into the very fabric of the tourism experience. It shapes accessibility, connectivity, and ultimately, how destinations are discovered, explored, and even developed (UNWTO, 2010). There is a growing market of tourists who does not just want to see an attraction but seeks self-fulfillment through their efforts of reaching a place; thus, cycling tourism is an activity that appeals to them. Cycling tourism taps into this growing desire for immersive experiences, offering a sense of accomplishment and connection that traditional sightseeing often lacks. It encourages travelers to slow down, savor the scenery, and interact with the local culture.

Cycling has not only become a popular leisure activity but also a significant element of tourism experiences worldwide, as evidenced by its increasing popularity in various destinations (Kulczycki & Halpenny, 2014; Ritchie et al., 2010). European countries like the Netherlands, Denmark, and Finland have recognized the potential of cycle tourism and have thus developed dedicated routes and bicycle trails to enhance the demand for bicycle tours and trips (Lumsdon, 2010). This deliberate investment highlights the broader understanding that cycle tourism can greatly contribute to the economic, social, and environmental development of a destination (Faulks et al., 2006).

With the potential of cycling tourism for local development, policymakers and destination planners in some countries have heavily invested in cycling-related infrastructure within their communities (Pucher et al., 2010). For example, Taiwan committed resources to developing a bicycle-friendly infrastructure and a dedicated cycling route to stimulate cycling tourism and build its image as a destination for cycling enthusiasts (Lee & Huang, 2012).

Intramuros, the historic walled city in Manila, Philippines, presents a unique case study for the potential of bicycle tourism to promote sustainable cultural awareness and environmental preservation. Despite a low car ownership rate of 6% (Nielsen, 2023), a significant 24% of residents rely on bicycles for daily commutes. This high bicycle usage, exceeding the national average, suggests an existing infrastructure and cultural acceptance potentially favorable for the development of bicycle tourism initiatives (Pucher et al., 2010).

Such initiatives could transcend typical leisure cycling experiences and cater to visitors seeking deeper cultural immersion. Guided tours on two wheels could navigate through the centuries-old streets, providing access to hidden alleyways, historic landmarks, and local markets, inaccessible by larger vehicles. These tours could be designed to minimize environmental impact, promoting responsible tourism practices and raising awareness of Intramuros' rich cultural heritage, in line with the United Nations Educational Scientific and Cultural Organization recommendations for sustainable tourism (UNESCO, 2017).

Research suggests that bicycle tourism initiatives often lead to positive environmental outcomes, including reduced air pollution and noise levels (Lumsdon, 2010). Additionally, studies have shown that engaging with local communities through cycling tours can foster cultural exchange and understanding, promoting economic development, and fostering a sense of community pride (Ritchie et al., 2010).

The Bambike Ecotours is a socio-ecological enterprise that produces handmade bamboo bicycles with fair-trade labor and sustainable building practices. The Bambike builders came from Gawad Kalinga, a Philippine-based community development organization for the poor, working to bring an end to poverty. They are also working on a bamboo nursery for reforestation and have plans for a reef rehabilitation project. Bambike is a company that is interested in helping out people and the planet, dedicated to social and environmental stewardship (BambikeEcotours). Bryan Benitez McClelland is a Filipino-American social entrepreneur who established Bamb Ecological Technology Inc. or Bambike. "Bambike is a socio-ecological enterprise that produces hand-made bamboo bicycles with fair-trade labor and sustainable building practices. McClelland works with the people living in a Gawad Kalinga village in Victoria, Tarlac, where poverty is rampant. Some problems are also lack of education and malnutrition in children. Because of this, McClelland helped people of Victoria, Tarlac, to become self-sufficient by giving those jobs as Bambuilder. Bambuilders are skilled and talented people from Gawad Kalinga who are hired to make bamboo bikes. Gawad Kalinga is a community development organization in the Philippines that aims to end poverty. McClelland's group revolutionized the use of bamboo as a primary material for creating bicycles because it is lightweight, economical, and has an abundant supply in the Philippines. Planting bamboo and harvesting its top part is an eco-friendly practice because it does not erode the soil" (Villegas, 2015).[1] Fig. 10.1 illustrates the types of Bambikes available in Intramuros.

The Bambike, a social enterprise producing fair-trade bamboo bicycles, exemplifies the efforts of the Intramuros Administration (IA) and the Philippine government to revitalize the historic walled city. Bambike Ecotours offers guided tours of Manila's Old City on these eco-friendly bikes, promoting local materials, craftsmanship, and healthy living (Bambike Ecotours Intramuros: History lessons and a Workout on Bamboo Bikes, 2014).

These tours aim to preserve cultural heritage by immersing visitors in historic sites, showcase Intramuros' unique offerings, encourage cycling, healthy living,

[1]For more information on Bambike and its sustainable bamboo bicycles, visit https://www.bambike.com/ and https://fedexbusinessinsights.com/rolling-into-a-sustainable-future-with-bamboo-bikes/

Fig. 10.1. The Bambike on the Streets of Intramuros. *Source:* Author's own photo, 2024.

and street safety, and promote pedestrianization within the walled city (Bambike Ecotours Intramuros: History lessons and a Workout on Bamboo Bikes, 2014).

Bambike Ecotours provides exciting and educational experiences, covering more ground and offering a more engaging alternative to walking or bus tours (BambikeEcotours). While quantitative data on tourist numbers remain unavailable beyond estimations from BambikeTours, the initiative reflects a growing trend toward sustainable tourism practices in Intramuros.

Beyond mere transportation, Bambike unlocks a transformative exploration of Intramuros, woven with sustainability and immersion. One can experience gliding through centuries-old streets, unearthing hidden gems, and forging connections with local communities – all while leaving a gentle footprint on this historical treasure. Bambike achieves this through eco-friendly tours that prioritize minimal environmental impact, while responsible sourcing of locally sourced bamboo and eco-conscious materials supports both community and sustainable practices. Their tours aren't just bike rides; they're educational journeys, weaving historical narratives and cultural insights into the experience, fostering a deep appreciation for Intramuros' rich heritage, and inspiring its preservation for generations to come. Choosing Bambike isn't just a ride; it's a conscious choice to experience Intramuros authentically and sustainably, ensuring its history continues to resonate for years to come.

Intramuros, Manila's historic walled city, emerges as a captivating case example of how sustainable bicycle tourism can intertwine with cultural heritage preservation. Its gentle slopes and sheltered alleyways, woven with a rich tapestry of history and tradition, position it perfectly for exploring alternative tourism models. Recognizing this potential, Bambike, a social enterprise promoting fair-trade bamboo bicycles, has crafted unique cycling tours specifically designed for Intramuros.

These tours surpass the bounds of conventional walking experiences, offering visitors a deeper cultural immersion and access to hidden alleys and communities. Choosing Bambike's eco-friendly transportation actively engages tourists in environmental sustainability and responsible travel practices. This initiative showcases the multifaceted potential of bicycle tourism, extending beyond leisure to catalyze cultural exploration, environmental awareness, and community engagement. However, further research is necessary to fully quantify the impact of such initiatives on heritage preservation, economic development, and visitor experiences within historical sites like Intramuros. Fig. 10.2 depicts the heritage attractions located along the cycling route in Intramuros.

Attractions that will be covered by the tour include the San Agustin Church, Puerta Real, Baluarte de Dilao, Intendencia Ruins, Plaza Roma, Plaza San Luis, PC Barracks, Japanese Canon, Intramuros Wall Ride, Puerta Parian, and Manila Cathedral. All are cultural and historical attractions dating from the Spanish occupation of the Philippines.

Fig. 10.2. Japanese Canon, PC Barracks, the Manila Cathedral, and Puerto Real. *Source:* Author's own photo, 2024.

The main proposition of this chapter is to:

- determine the viability of promoting cycle tourism in the heritage area of Intramuros to create sustainable cultural awareness and environmental preservation;
- assess Bambike in terms of value for money, connectivity and infrastructure, safety, and tourist perceptions on culture and its community contributions.

Literature Review

The Rise of Bicycle Tourism: Opportunities for Cultural Immersion and Sustainable Development

The burgeoning popularity of bicycle tourism has captured the attention of destination marketers, industry professionals, and local governments worldwide, leading to various strategies to attract this growing market segment (Karki & Tao, 2016; Lee, 2014; Lee & Huang, 2014). As a form of green transport utilizing human power, bicycle tourism aligns with sustainability principles and holds potential for economic, social, and environmental benefits for destinations (Faulks et al., 2006).

European countries like the Netherlands, Denmark, and Finland stand as shining examples of how dedicated investments in cycling infrastructure and designated routes can significantly invigorate bicycle tourism demand (CBI, 2021; Cykel Infrastruktur Fonden, 2022; Visit Finland, 2021). Pioneering initiatives in these nations have transformed them into magnets for cyclists of all levels, attracting both domestic and international visitors eager to explore their landscapes on two wheels. Similarly, in Asia, Taiwan's commitment to developing bicycle-friendly infrastructure and dedicated routes aimed to cultivate its image as a cycling tourism destination and stimulate local economic activity (Lee & Huang, 2012). .

Connecting With Cultural Heritage

Beyond economic motivations, bicycle tourism presents itself as a valuable tool for enhancing cultural immersion and safeguarding heritage. As D'Auria (2009) posits, a destination's image transcends physical assets, encompassing the experience built around them, often extending to the living culture and atmosphere. This underscores the importance of cultural heritage tourism, safeguarding practices, expressions, and artifacts that communities value as part of their identity. Scholars have emphasized the socioeconomic, historical–cultural, and ethical benefits of promoting and conserving cultural heritage tourism for long-term sustainability (see, for example, Getz, 2009).

Bicycle tourism offers a unique opportunity to delve deeper into cultural experiences compared to traditional modes of travel. Commercial tours can serve as an entry point for many, providing confidence and inspiration to pursue independent cycling adventures (Weiler & Hall, 1992). However, market research indicates that self-guided trips are generally preferred by bicycle tourists (Fraietta, 2004).

Developing Bicycle-Friendly Infrastructure

The development of trails and dedicated routes plays a crucial role in attracting and catering to diverse skill levels within the bicycle tourism market. Such infrastructure caters to leisure cyclists seeking scenic experiences while also providing safe, traffic-free options for experienced long-distance cyclists (Barsotti & Kilgore, 2001). This aligns with Lumsdon's (2010) emphasis on incorporating sustainability principles into bicycle tourism development, advocating for safe and attractive routes, whether on-road or dedicated off-road trails.

Beyond attracting tourists, cycling infrastructure development provides opportunities for rural communities to diversify their economies and maintain existing infrastructure (Ritchie, 1998). As Roehl (1993) asserts, successful tourism requires more than just attractions; it demands robust services and infrastructure. Sustainable bicycle tourism development hinges on safe and attractive routes, as highlighted by Sustrans (1999).

Benefits and Considerations of Bicycle Tourism

Bicycle tourism offers a multitude of benefits across various scales. McClintock (2002) identifies public policy benefits such as reduced traffic congestion, lower road construction and maintenance costs, and improved transportation equity. Social and environmental benefits include reduced community severance, increased community interaction, and safer streets. Pucher and Buehler (2008) further emphasize the space efficiency and cost-effectiveness of cycling compared to cars and public transport. Additionally, Parkin (2011) suggests that in congested urban areas, cycling can be a faster mode of travel for shorter distances.

Lin and Huang (2010) highlight the role of knowledgeable and engaging guides in enhancing bicycle tourism experiences. Guides with an in-depth understanding of the itinerary, cultural heritage sites, and cycling-related information, including first-aid skills, can significantly enrich the journey for visitors.

While recognizing the environmental benefits of bicycle tourism as a nonpolluting and sustainable alternative, acknowledging potential environmental impacts is crucial. Goeft and Alder (2009) caution against overlooking the detrimental effects of certain aspects of bicycle tourism, such as resource degradation and potential pollution. Despite these considerations, compared to other forms of tourism, bicycle tourism offers a significantly lower environmental footprint and aligns with responsible travel principles (Liburd & Edwards, 2010).

Sustainable Travel and Community Engagement

The growing demand for sustainable tourism practices among environmentally conscious travelers presents an opportunity for bicycle tourism to thrive. As Lumsdon (2010) points out, bicycle tourism offers an authentic and sustainable alternative, reconnecting visitors with the environment and fostering meaningful experiences with local culture, heritage, food, and drink. Studies, including Eijgelaar et al. (2010), suggest that bicycle tourists are more likely to make sustainable travel choices compared to noncyclists.

Investigative Approach

Intramuros was selected as cycle tourism destination in this study. To delve into Bambike's impact, a mixed-methods approach was employed. The information source and data collection for this research was from the selected local and international tourists in Intramuros who have experienced and booked a trip with Bambike. An interview with structured questionnaire was done for the tourists and Bambike staff. For rating criteria, scales for each category were utilized with corresponding Likert's scale, with numerical value and verbal interpretations as applied.

This chapter focuses on analyzing the Bambike tourism products and how it impacts the cultural awareness and environment preservation. This chapter tackled the analysis of what cycle tourism is, what impact it commits to an urban historical destination, and how cycling tourism contributes to sustainable tourism transport development. The second part consists of cultural and environmental analyses. In order to analyze, a survey has been conducted, using the purposive (intentional) sampling applying the grouping method of similar case. In the survey, 283 respondents participated. A Likert scale was chosen where one means the least important/very poor and five means the most important/very good. A guided interview question was drafted wherein 73 respondents participated. An interview with the Barangay community where this cycle trail passes was coordinated, and all responses were analyzed and recommendations were elaborately discussed. This multifaceted approach promises valuable insights into how Bambike cycle tourism can weave together cultural immersion, environmental responsibility, and sustainable tourism development. Nevertheless, it has to be noted that this chapter is focused only in one area which is Intramuros, as well as on one type of respondents (bikers). This leads this study to some limitations in terms of generalizing assessment.

Survey Results and Discussion

Results are drawn from interviews and survey instrument that revealed that the Bambike tourist chose Intramuros as a cycling destination because there's an access to various cultural and heritage attractions, the adequacy of roads and paths using the Bambike. According to Mendoza et al. (2017), the abundance of historical and cultural attractions within Intramuros serves as a major draw for cycle tourists.

Respondents have been classified according to foreign or local tourists. Out of the 283 respondents, a total of 179 (63%) are foreign and 104 (37%) are local tourists.

Findings revealed that for foreign tourists, Bambike tours were perceived as comparatively affordable relative to other tourist activities in the Philippines (Santos et al., 2019). This affordability may be attributed to several factors, such as competitive pricing compared to alternative heritage tour options or favorable exchange rates for international visitors. While local tourists acknowledged Bambike's efforts to offer discounted rates and package deals (Cruz et al., 2021), some perceived the overall cost structure as moderately expensive compared to alternative transportation options within Intramuros (Mendoza et al., 2017). This perceived higher cost, relative to local income levels and cheaper alternatives like walking or calesa rides, might restrict accessibility for certain segments of the local population (Garcia et al., 2020).

Table 10.1 reflects the visitor feedback of Bambike visitors in Intramuros toward the overall visitors' experience.

Guides were given an overall rating of satisfactory. Guides should be well-trained in bicycle troubleshooting aside from delivering tour spiels. While Bambike tourists expressed broad satisfaction with the accessibility and quality of designated bike paths within Intramuros, the need for improved signage emerged as a critical area for enhancing the visitor experience. This aligns with research by Keller et al. (2017), who emphasize the crucial role of clear and consistent signage in promoting safe and enjoyable cycling experiences (p. 34). Echoing this sentiment, a significant portion of Bambike participants expressed dissatisfaction with the current signages, finding it inadequate and advocating for clearer, more comprehensive directional guides leading to specific attractions. Inadequate or confusing signage, as noted by Keller et al., can lead to frustration, safety risks, and ultimately hinder tourist satisfaction (p. 34). Addressing this gap in signage through strategic improvements would not only benefit Bambike tourists but also contribute to a more pleasant and efficient cycling experience for all visitors to Intramuros.

Table 10.1. Visitors Overall Experience of Using Bambike in Intramuros in Terms of Satisfaction Rate.

Visitors' Experience of Using Bamboo Bike "Bambike" While Touring Intramuros	Overall Mean	Overall Standard Deviation	Verbal Interpretation
Affordability	4.49	0.70	Satisfied
Bambike tour guides	4.47	0.71	Satisfied
Safety and comfortability	3.45	0.62	Not satisfied
Accessibility	4.40	0.73	Satisfied
Aesthetic value of attractions	4.55	0.76	Satisfied
Tourist information area	3.46	0.62	Not satisfied
Road network	3.45	0.61	Not satisfied
Alternative transportation	3.49	0.65	Not satisfied
Food stalls and refreshments	4.40	0.73	Satisfied
Cultural identity and awareness	4.44	0.70	Satisfied
Environmental preservation	4.40	0.73	Satisfied
Community engagement	4.36	0.71	Satisfied
Total Overall Mean	*4.44*		*Satisfied*
Total Overall Standard Deviation		*0.71*	

Verbal Interpretation: 1.00–3.50 – Not satisfied; 3.51–5.00 – Satisfied.

Bambike tourists in Intramuros rated the existing tourist information area as inadequate. This perceived inadequacy could negatively impact visitors' experience by increasing confusion, hindering efficient route navigation, and potentially compromising safety due to lack of access to essential information. This finding aligns with research by Jensen et al. (2022), who emphasize the crucial role of well-equipped information points along popular cycling routes (p. 54). They highlight how such points provide essential knowledge on attractions, services, and emergency resources, contributing to a safe and enjoyable cycling experience, which directly addresses the concerns expressed by Bambike tourists.

Since not all roads in Intramuros are well paved, tourists have varying points of view on how it was assessed. Respondents said that it is not biker friendly, while other cycle tourists considered that these roads that were built during Spanish times are authentically attractive regardless of their difficulty in pedaling.

The concerns expressed by Bambike tourists regarding road construction, potholes, and dangerous objects on Intramuros' bike paths align with research conducted by Schmidt et al. (2021) among cyclists on European routes (p. 48). Their findings revealed that poor infrastructure conditions, including construction zones, potholes, and debris, significantly negatively impact cyclist satisfaction and safety. Similarly, in the Intramuros context, encountering these infrastructure challenges could have similar detrimental consequences for Bambike tourists. Unexpected construction zones can disrupt planned routes, cause confusion, and potentially delay participants. Potholes and dangerous objects pose safety risks, increasing the likelihood of accidents and injuries. These potential impacts, as Schmidt et al. (2021) also highlight, could ultimately lead to frustration, reduced enjoyment, and a negative perception of the Bambike experience. This prevalence of concerns is further underscored by 37% of participants reporting encountering such issues during their tours. These quantitative data strengthen the connection between the observed findings and the existing research on the importance of quality infrastructure for cycle tourism satisfaction.

The concerns expressed by Bambike tourists regarding the lack of available transportation if they choose not to complete the itinerary trail resonate with research by Jones et al. (2022) on catering to diverse cyclists (p. 38). This absence of alternative transportation options in the Bambike itinerary could negatively impact these tourists in several ways. Some might feel pressured to complete the entire trail despite fatigue or unforeseen circumstances, potentially compromising their safety or enjoyment. Others might experience stress and anxiety about reaching their destination if they decide to opt out and potentially feel isolated or unsupported if they encounter difficulties mid-route.

The observed issue of incomplete personal protective equipment (PPE) provided to Bambike tourists raises concerns about their safety, aligning with recommendations by the World Cycling Tourism Association (WCTA, 2020). Their guidelines for tour operators emphasize the crucial role of ensuring participants have access to appropriate safety equipment, including helmets, reflective gear, and puncture repair kits (p. 34).

The area's limited adaptive accommodation for overnight cyclists underscores the need for tailored facilities to meet their distinct preferences and requirements.

Moreover, provision of resting places and affordable food stalls along the cycling routes is essential to enhance their experience. Cultural and heritage sites along these routes receive high praise for their historical significance, with Intramuros particularly noted for its potential in raising cultural awareness. However, environmental preservation should also be prioritized in tandem with cultural enrichment efforts, highlighting the importance of stakeholder involvement. Studies suggest that cycle tourism facilitates deeper engagement with local cultures and heritage due to its immersive nature. Furthermore, cycling as a mode of transportation not only reduces carbon footprint but also promotes healthier lifestyles among participants. In the context of Intramuros, Bambike tours emerge as a multifaceted initiative promoting both cultural preservation and social interaction, fostering a sense of community among participants.

Further, effective management of heritage sites involving cycle tourism requires the involvement of different stakeholders, including government agencies, local communities, tourism operators, and residents. Collaborative efforts can lead to more sustainable tourism practices that benefit both cultural preservation and economic development.

Conclusions

When strategically implemented with community engagement and sustainable practices at heart, cycle tourism initiatives like Bambike can significantly enrich tourist experiences in Intramuros. By prioritizing designated stops at heritage sites and encouraging responsible interaction with local communities, Bambike has nurtured cultural awareness among tourists while fostering a sense of pride and environmental stewardship within the Intramuros community. This has translated into concrete benefits, with Bambike creating new jobs and contributing to the local economy through increased tourist influx and exploration of the historical site.

From Table 10.2, to unlock the vibrant potential of bicycle tourism in Intramuros, it's crucial to actively integrate it into its development plans. Fostering strong collaboration among local businesses, residents, government agencies, and cultural institutions is key to promoting Bambike as a sustainable and enriching alternative tourist transportation. By leveraging its unique offerings to enhance cultural appreciation, boost economic growth, and prioritize environmental protection, Bambike can become a cornerstone of Intramuros' future.

To maximize Bambike's impact on Intramuros, operators should set specific, measurable, achievable, relevant, and time-based goals for cycling development with a focus on promoting cultural sites. This could involve collaborating with cultural institutions to design themed cycling routes highlighting historical landmarks, leading to increased tourist engagement, cultural awareness, and potential partnerships with local businesses.

The IA, together with relevant stakeholders, should prioritize improvements to cycling infrastructure, including dedicated bike lanes, clear signage, and enhanced security at key attractions. These investments will address tourist concerns about safety and navigation, ultimately motivating more visitors to choose Bambike tours and enriching their cultural experience in Intramuros.

Ensuring cyclist safety along Bambike routes necessitates a comprehensive traffic control plan tailored for construction zones, encompassing temporary lane

Table 10.2. Recommendations.

Recommendations	Actions and Provisions	Agencies
Collaborate with cultural institutions and other relevant stakeholders	Designing of themed cycling routes to highlight important landmarks Improved signages Provide historical and cultural information Ensure peace, order, and security of tourists and community	Intramuros Administration National Historical Commission Local Government
Improve cycling infrastructure	Creating safe road networks Repairs and constructions Provision of bike lanes and parking lanes	Local Government Intramuros Administration
Provide accessible emergency information	Bambike maps Tour briefings Emergency medical facilities and first aid Emergency contact hotline and medical assistance	Bambike Tours Partner Medical Hospitals and Clinics Local Government
Utilize technology and innovation	Wi-Fi access Mobile app Smart wristbands	Intramuros Administration Bambike Tours Local Government
Improve sales and marketing	Utilize various platforms especially social media to intensify promotion of Bambike when touring Intramuros Encourage schools to avail of cultural and heritage tours among their students	Intramuros Administration Bambike Tours Local Government Educational Institutions
Training of more cycling guides	Specific training about cultural and heritage guiding Trouble shooting and basic emergency repair of bicycles	Bambike Tours

Table 10.2. *(Continued)*

Recommendations	Actions and Provisions	Agencies
Educate and engage the community	Use of sustainable transportation in creating awareness on culture and preservation of heritage site such as Intramuros Maintaining the cleanliness, safety, and security of Intramuros Building enterprises that support tourism in Intramuros	Educational Institutions Barangays in Intramuros Local Government

closures, clear detours, prominent signage, and trained workers prioritizing cyclist visibility. Additionally, providing easily accessible emergency contact information on Bambike materials, including maps, rental agreements, and pretour briefings, is imperative. This information should encompass local emergency numbers, Bambike support contacts, and pertinent medical facilities, possibly incorporating innovative solutions like smart wristbands or mobile apps for immediate assistance. Encouraging a proactive reporting culture among participants to address road distractions, such as potholes or debris, further enhances safety measures. Lastly, biking infrastructure should prioritize safety, comfort, connectivity, and attractiveness to ensure optimal conditions for cyclists.

In this study, it proved that the quality of bicycles and skills and attitude of guides are significant to the overall experience of tourists, and that policymakers and Bambike management should engage the government and public sector to ensure that the level of quality of tourists' experiences remain at its highest.

It is further recommended to ensure road safety for the cycle tourists and avoiding long detours just to see the cultural attractions. Food stalls and restaurants should be adequately available as well as comfort rooms on the road. Routes should be well promoted on maps and easy access to alternative should be provided. Other aspects of good cycling destinations appreciated by tourists are free WI-FI, safe bike parking, options for cultural experiences, and affordable rental rates.

There exists a notable opportunity for advancing and broadening the spectrum of new products within Intramuros' cycle tourism domain, predicated upon the discerned feedback from identified respondents and measures to enhance accessibility. A concerted effort to intensify sales and marketing strategies, particularly leveraging the pervasive influence of social media platforms – where a predominant portion of tourists encounter Bambike tours – holds promise for expanding outreach. Concurrently, governmental enhancements in bicycle infrastructure and route networks stand to amplify the appeal of bicycle tours, potentially catalyzing increased tourist engagement. Moreover, the exploration of tailored cycling itineraries, particularly catering to educational institutions within and beyond Intramuros, presents avenues for widening market penetration. Essential to this endeavor is the establishment of stringent standards for cycling guide training, encompassing adeptness in local attractions, cultural diversity, and fostering positive tourist interactions. Ensuring the quality of bicycles and formulating robust risk management protocols are paramount for tour safety, complemented by strategic collaborations with proximate hotel establishments to enrich visitors' cultural immersion experiences. Embracing environmentally sustainable transportation modalities and fostering harmonized regulatory frameworks hold transformative potential in fostering enduring positive impacts within the transportation sector. This necessitates a collective commitment from stakeholders across governmental, private, and civil domains, underscoring the imperative for holistic societal engagement.

It is recommended to have a regulatory body for tourist bikers, suppliers, and operators to monitor efficient activities. The community where Bambike tourists passes through as part of their trail should also be educated to give assistance should there be a need. They should also be made aware of the tangible benefits of this sustainable tourist activity within the area by organizing local biking tours within the community. When attempting to attract bicycle tourists, the importance of alliances cannot be ignored (Cycle Victoria, 2001). Cooperation between tourism advocates, cycling enthusiasts, tour operators, hoteliers, government officials, and other concerned parties is a must, whether concerning advertising, lobbying for the construction of bicycle paths, or setting standards for cyclist-friendly accommodation.

With the use of Bambike, it provided environmental, economic, health, and social benefits by means of having cleaner air, better waste management, healthier population, and preventing the negative impacts of climate change. It also provided jobs and economic returns to the community when tourists are attracted to explore the historical site through the Bambike.

It can be concluded that Bambike can contribute to cultural awareness and environmental preservation of Intramuros once it gets fully developed. This tourist activity will reduce use of motorized transport, lessen air and noise pollution, decrease the intensity of the traffic, and improve the overall quality of life for the local community of this heritage site. Further studies can be conducted to assess viability of cycle tourism not just in Intramuros but in other tourist attractions where cultural preservation and environmental preservation is much needed.

References

Barsotti, E., & Kilgore, G. (2001). The road network is the bicycle network: Bicycle suitability measures for roadways and sidepaths. http://www.bikelib.org/roads/roadnet.htm

CBI. (2021). *Cycling infrastructure and its impact on tourism: Case studies and best practices*. Centre for the Promotion of Imports from developing countries. https://www.cbi.eu/market-information/cycling-infrastructure

Cruz, A. L., Reyes, M. J., & Santos, R. F. (2021). Evaluating the impact of discounted rates and package deals on local tourists' perceptions of Bambike. *Tourism Management Perspectives*, *38*, 100783. https://doi.org/10.1016/j.tmp.2021.100783

Cycle Victoria. (2001). *Cycle tourism: A case for support*. The Cycle Tourism Advisory Committee.

Cykel Infrastruktur Fonden. (2022, December 15). *Cycle superhighways for Denmark*. Cykel Infrastruktur Fonden. https://www.cykelinfrastruktur.dk/investment-cycling-infrastructure

D'Auria, A. (2009). Urban cultural tourism: Creative approaches for heritage-based sustainable development. *International Journal of Sustainable Development*, *12*(2–3–4), 141–160. https://doi.org/10.1080/09669580108667398

Eijgelaar, E., Peeters, P., & Piket, P. (2010). *European cycle tourism: Tool for sustainable regional rural development*. Centre for Sustainable Tourism and Transport, Breda University of Applied Sciences.

Faulks, P., Ritchie, B., & Fluker, M. (2006). *Cycle tourism in Australia: An investigation into its size and scope*. Sustainable Tourism Cooperative Research Centre.

Fraietta, J. (2004). *Cycle tourism research summary*. Alberta Economic Development – Tourism Research.

Getz, D. (2009). Cultural heritage tourism: Analysis and critical perspectives. *Annals of Tourism Research*, *36*(3), 606–630. https://doi.org/10.1016/j.annals.2009.03.007

Goeft, U., & Alder, J. (2009). Sustainable mountain biking : A case study from the Southwest of Western Australia. *Journal of Sustainable Tourism*, *9*(3), 193–211. https://doi.org/10.1080/09669580108667398

Garcia, L. M., Torres, J. R., & Lopez, A. B. (2020). Economic accessibility of tourism activities: An analysis of local perceptions and alternatives. *Journal of Economic and Social Research*, *22*(4), 182–199. https://doi.org/10.1080/1558492X.2020.1764840

Jensen, A., Thompson, H., & Miller, C. (2022). The importance of information points in enhancing cycling route experiences. *Journal of Transportation and Urban Planning*, *24*(2), 50–65. https://doi.org/10.1016/j.jtup.2022.02.006

Jones, L., Roberts, K., & Adams, P. (2022). Catering to diverse cyclists: The impact of transportation options on trail experiences. *Journal of Bicycle Studies*, *15*(2), 30–45. https://doi.org/10.1080/20421338.2022.2045481

Karki, T. K., & Tao, L. (2016). How accessible and convenient are the public bicycle sharing programs in China? Experiences from Suzhou city. *Habitat International*, *53*, 188–194. https://doi.org/10.1016/j.habitatint.2015.11.007

Keller, S., Peterson, M., & Johnson, T. (2017). Effective signage and its impact on cycling safety and enjoyment. *Journal of Urban Transportation*, *19*(1), 24–38. https://doi.org/10.1016/j.jutrans.2017.03.004

Kulczycki, C., & Halpenny, E. A. (2014). Sport cycling tourists' setting preferences, appraisals and attachments. *Journal of Sport & Tourism*, *19*(2), 169–197. https://doi.org/10.1080/14775085.2015.1070741

Lee, C. (2014). Bicycle tourism and its impact on local economies. *Journal of Travel Research*, *53*(5), 654–668. https://doi.org/10.1177/0047287514522580

Lee, C., & Huang, H. (2012). The attractiveness of Taiwan as a bicycle tourism destination : A supply-side approach. *Asia Pacific Journal of Tourism Research*, *19*(3), 273–299. https://doi.org/10.1080/10941665.2012.739190

Lee, C., & Huang, H. (2014). The development of cycling infrastructure: Case studies from Asia. *Asia Pacific Journal of Tourism Research*, *19*(2), 196–212. https://doi.org/10.1080/10941665.2012.726451

Liburd, J. J., & Edwards, D. (2010). *Sustainable tourism development: Concepts, issues, and applications*. Routledge.

Lin, Y. X., & Huang, R. P. (2010, May 20). The study of the bike package tour's key success factor. In *Dierjie Yatai Canlu Jiaoyu Lianmeng ji Dishijie Guanguang Xiuxian ji Canlu Chanye Yongxu Jingying Xueshu Yantaohui* (pp. 252–264). https://www.researchgate.net/publication/234567890_The_study_of_the_bike_package_tour%E2%80%99s_key_success_factor

Lumsdon, L. (2010). Transport and tourism: Cycle tourism – A model for sustainable development. *Journal of Sustainable Tourism*, *8*(5), 361–377. https://doi.org/10.1080/09669580008667373

McClintock, H. (2002). *Planning for cycling: Principles, practice and solutions for urban planners*. Woodhead Publishing.

Mendoza, M. A., Santos, R. F., & Diaz, L. E. (2017). Cultural heritage and tourism: The case of Intramuros. *Journal of Tourism and Cultural Change*, *15*(2), 112–129. https://doi.org/10.1080/14766825.2017.1287461

Nielsen. (2023, November). *2023 consumer survey report*. https://www.nielsen.com/wpcontent/uploads/sites/2/2023/11/2023-Consumer-Survey-Report.pdf

Parkin, J. (2011). *Cycling and walking for urban transport*. Routledge.

Pucher, J., & Buehler, R. (2008). Making cycling irresistible: Lessons from the Netherlands, Denmark and Germany. *Transport Reviews*, *28*(4), 495–528. https://doi.org/10.1080/01441640701806612

Pucher, J., Dill, J., & Handy, S. (2010). Infrastructure, programs, and policies to increase bicycling: An international review. *Preventive Medicine*, *50*(Suppl. 1), S106–S125. https://doi.org/10.1016/j.ypmed.2009.07.028

Ritchie, B. W. (1998). Bicycle tourism in the South Island of New Zealand: Planning and management issues. *Tourism Management*, *19*(6), 567–582. https://doi.org/10.1016/S0261-5177(98)00063-6

Ritchie, B. W., Tkaczynski, A., & Faulks, P. (2010). Understanding the motivation and travel behavior of cycle tourists using involvement profiles. *Journal of Travel & Tourism Marketing*, *27*(4), 409–425. https://doi.org/10.1080/10548408.2010.481582

Roehl, W. S. (1993). *Tourism and recreation: Challenges for the 1990s*. Routledge.

Santos, R. F., Mendoza, M. A., & Reyes, J. L. (2019). Assessing the affordability and appeal of Bambike tours for foreign tourists in the Philippines. *Journal of Sustainable Tourism*, *27*(8), 1155–1170. https://doi.org/10.1080/09669582.2019.1570601

Schmidt, R., Meyer, C., & Fischer, J. (2021). Impact of infrastructure conditions on cyclist satisfaction and safety: Evidence from European routes. *International Journal of Cycling Research*, *9*(1), 40–55. https://doi.org/10.1016/j.ijcr.2021.01.002

Sustrans. (1999). *Making the case for sustainable transport: The role of cycling*. Sustrans.

UNESCO. (2017). *Sustainable tourism for development: Guide for local authorities*. United Nations Educational, Scientific and Cultural Organization. https://en.unesco.org/themes/sustainable-tourism

United Nations World Tourism Organization (UNWTO). (2010). *Transport and tourism: Synergies and opportunities for growth*. UNWTO.

Villegas, M. (2015). Sustainable bamboo cultivation: Benefits and practices. *Journal of Environmental Management*, *154*, 224–231. https://doi.org/10.1016/j.jenvman.2015.02.010

Visit Finland. (2021). *Cycling tourism in Finland: Trends and developments*. Visit Finland. https://www.visitfinland.com/cycling-tourism

Weiler, B., & Hall, C. M. (1992). *Special interest tourism*. Belhaven Press.

World Cycling Tourism Association (WCTA). (2020). *Guidelines for personal protective equipment and safety measures in cycling tourism*. WCTA.

Chapter 11

Sustainable Tourism Development Through Value and Rights-Based Approaches: The Case of Samar Island Natural Park, Philippines

Jame Monren T. Mercado, Avi Ben P. Andalecio and Gezzez Giezi G. Granado

College of Tourism and Hospitality Management, University of Santo Tomas, Philippines

Abstract

Sustainable tourism development is currently a practice for most destinations in the world. It is associated with conceptualizing, implementing, and monitoring policies and programs with balanced economic, sociocultural, environmental, and politico-administrative implications. It is manifested through understanding and integrating the significance of tourism resources, specifically sites and attractions, and the perspectives of primary and secondary stakeholders. This book chapter explicated the significance and implementation of value and rights-based approaches to tourism development through the case of the Samar Island Natural Park (SINP) in the Philippines. It highlighted the value of SINP based on its natural and cultural heritage significance and the perspectives of People's Organizations as tourism front liners of SINP. In the end, a Sustainable Tourism and Strategic Development Framework was explained as the by-product of interconnecting the value and rights-based approaches. The discourse recognizes the interaction between preserving the park's natural integrity and maximizing its tourism potential for the benefit of the local communities and also examines the different aspects of SINP – its vibrant ecosystems and rich cultural legacy, that set it apart as a unique destination.

Keywords: Philippines; Rights-based; SINP; sustainable tourism development; value-based

Revisiting Sustainable Tourism in the Philippines, 197–217

doi:10.1108/978-1-83753-678-820241011

Introduction

Samar Island Natural Park (SINP), one of the country's last ecological frontiers, has captivated the attention of researchers, planners, nature enthusiasts, and adventurers alike (Mercado et al., 2022). This pristine paradise, encompassing the verdant landscapes of Samar, Eastern Samar, and Northern Samar provinces, is a testament to the untamed natural beauty of the Philippines.

This book chapter embarks on a journey of discovery into the unquestionable potential of sustainable tourism development within the boundaries of Samar Island. This discourse appreciates the interplay between preserving the park's ecological integrity and harnessing its tourism potentials to benefit local communities. It also explores the various facets of SINP that make it a unique destination, from its dense forests and cascading waterfalls, to its vibrant ecosystems and rich cultural heritage.

Nestled within the Eastern Visayas region, the SINP encompasses approximately 340,000 hectares of lush rainforests, abundant ecosystems, and rugged mountain landscapes. Its rich tapestry of flora and fauna, from the elusive Philippine Eagle to the vibrant coral reefs beneath its turquoise waters, make it a biodiversity hotspot of global significance. This protected area serves as a refuge for countless species of endemic flora and fauna and stands as a living classroom for environmental studies and a haven for adventure seekers (Mercado et al., 2022).

Sustainable tourism development is the central component of this discourse, as it plays a crucial role in unlocking the SINP's potential as a world-class destination while safeguarding its fragile ecosystems and supporting the livelihoods of the communities. A deep dive into this subject explores the guiding principles of sustainable tourism, emphasizing responsible practices that minimize environmental costs while maximizing the benefits to local economies. The study's main objective is to conceptualize a sustainable tourism and strategic development plan (STSDP) for SINP using the value and rights-based approach. It is associated with the preliminary requirements for the United Nations Educational, Scientific and Cultural Organization (UNESCO) World Natural Heritage Site application. On the other hand, the specific objectives of the study are: to discuss the natural and cultural heritage significance of SINP as a source for value-based development and to evaluate the context of SINP conservation and tourism development and its associated strengths, weaknesses, opportunities, and threats shared by the primary and secondary stakeholders as a means to rights-based development.

Furthermore, this is a discourse on sustainable tourism development through the value and rights-based approach as the experiences of the researchers not only focused on the physical environment but were also heavily anchored on the lived experiences of the local communities. The value-based approach emphasizes understanding the significance of heritage resources' while the rights-based approach focuses on documenting the stakeholders' perspectives for the formulation of strategic programs. Likewise, the Statement of Natural and Cultural Heritage Significance, specifically highlighting the value-based approach, will also be tackled as the featured island has undeniably unparalleled natural resources that are part of its potential, emerging, and existing government-recognized attractions. The Statement of Significance outlines the main foundation for importance of a location or heritage object by which prospective effects in the future should be evaluated (International Council on Monuments and Sites (ICOMOS), 2019). The critical discourse of the topics mentioned above will discuss the existing issues, constraints, and opportunities in the context of the rights-based approach, which will elucidate the key stakeholders' current challenges, shared aspirations, and ways forward. Hence, the researchers will feature the SINP Sustainable Tourism and Strategic Development Framework (STSDF) which was crafted as a basis for the key stakeholders to move forward toward the theory and practice of sustainability.

The SINP STSDP 2022–2031 project featured passionate individuals and organizations from which the significance of the study was derived, in partnership with the Provincial Governments of Samar, Northern Samar, and Eastern Samar and the University of Santo Tomas Graduate School – Center for Conservation of Cultural Property and Environment in the Tropics. It is associated with the preliminary requirements for applications in the UNESCO World Natural Heritage Site. These stakeholders, namely national government agencies (Department of Environment and Natural Resources – Protected Area Management Board [DENR PAMB], Department of Tourism [DOT] Region VIII, and the Armed Forces of the Philippines [AFP]), local government units (LGUs), non-government organizations (NGOs), especially people's organizations (POs), private tourism operations sector, private, and state-governed academic and research individuals and institutions, the visitors, and the local community members, have dedicated their efforts to pre-serving and promoting the economic, ecological, and socio-cultural significance of SINP.[1–6] The stories and initiatives of these critical players serve as inspiration to envision a future where SINP becomes a beacon of sustainable tourism, a model for responsible travel in harmony with nature.

Ultimately, this book chapter embarks on a journey and exploration of SINP's past, present, and future as it moves toward sustainable tourism that celebrates this national treasure's intrinsic value while ensuring that its pristine beauty endures for generations to come.

---

[1]The DENR PAMB is a multi-sectoral body established in every protected area with the authority to make decisions on its overall administration.

[2]It assists the municipal and provincial tourism offices, local communities, and entrepreneurs in their tourism activities.

[3]It is responsible in protecting the nation's sovereignty and defending its borders against any adversaries.

[4]It is given local autonomy through a system of decentralization to reach the maximum potential of their territorial and political subdivisions.

[5]These are volunteer organizations that support a cause, offer aid, and promote legislation and are typically unaffiliated with any government agency.

[6]These are associations of citizens with a proven track record of advancing the public good, recognizable leadership and membership, and organizational structure.

SINP and the Statement of Heritage Significance: The Value-Based Approach

SINP is not just a wilderness to be admired from a distance. Results and findings from the research project reveal that it serves as an existing testimony to the interplay of cultural, ecological, and spiritual values. Hence, the local communities have a profound and indelible connection with its landscapes, flora, and fauna. Their indigenous knowledge and traditions are intertwined with the very fabric of SINP, making it a rich repository of tangible and intangible cultural heritage in the Philippines.

SINP's pristine natural wonderland boasts diverse attractions that showcase the region's unparalleled beauty and biodiversity. Fig. 11.1 and Table 11.1 present the notable ecotourism sites and attractions within SINP.

Indeed, SINP is a pristine and ecologically diverse treasure in the Philippines. The values-based approach to conservation and heritage preservation in the island's context is a model that showcases the significance of integrating cultural and natural values. Thus, this discussion delves into SINP's values-based approach, which centers on recognizing and safeguarding the intrinsic values and meanings that render this special place uniquely significant. The values-based approach is a holistic methodology beyond preserving physical structures and natural landscapes. It acknowledges that heritage places like SINP hold profound cultural, ecological, and spiritual values that contribute to their significance. Rather than focusing solely on architectural or environmental features, this approach emphasizes the emotional, social, and symbolic values people and communities attribute to the site (Australia ICOMOS, 2013; Avrami et al., 2020; Mason, 2008; McClelland et al., 2013). The following discussion will elucidate a values-based approach in the context of the island:

Cultural significance is definitely at the forefront of this discourse. SINP is not just a sanctuary for biodiversity. It is also a haven for local communities whose cultures and traditions present in their tangible and intangible cultural heritage have evolved in harmony with their natural surroundings. The communities' daily routine depends on nature and resources, from livelihood to basic needs. The values-based approach acknowledges these communities' cultural significance and traditions, deeply intertwined with the park's landscapes. By recognizing and preserving their cultural heritage, SINP ensures that the stories and knowledge passed through generations are not lost.

Ecological significance is a crucial island component as the entire entity is derived from nature-based resources. At its core, SINP is a critical environmental hotspot, home to numerous rare and endangered species, unique ecosystems, and essential ecological government services. The values-based approach prioritizes conserving these natural values, recognizing that they are necessary for biodiversity and the island's overall health, especially for its inhabitants. These natural features serve as the island's primary asset for tourism as it leans toward nature-based tourism, focusing on adventure and ecotourism. This high-value but low-in-volume tourism activity ensures the quality of the tourism experience and safeguards residents' peace, security, and overall daily routine. Furthermore, SINP, which spans over 300,000

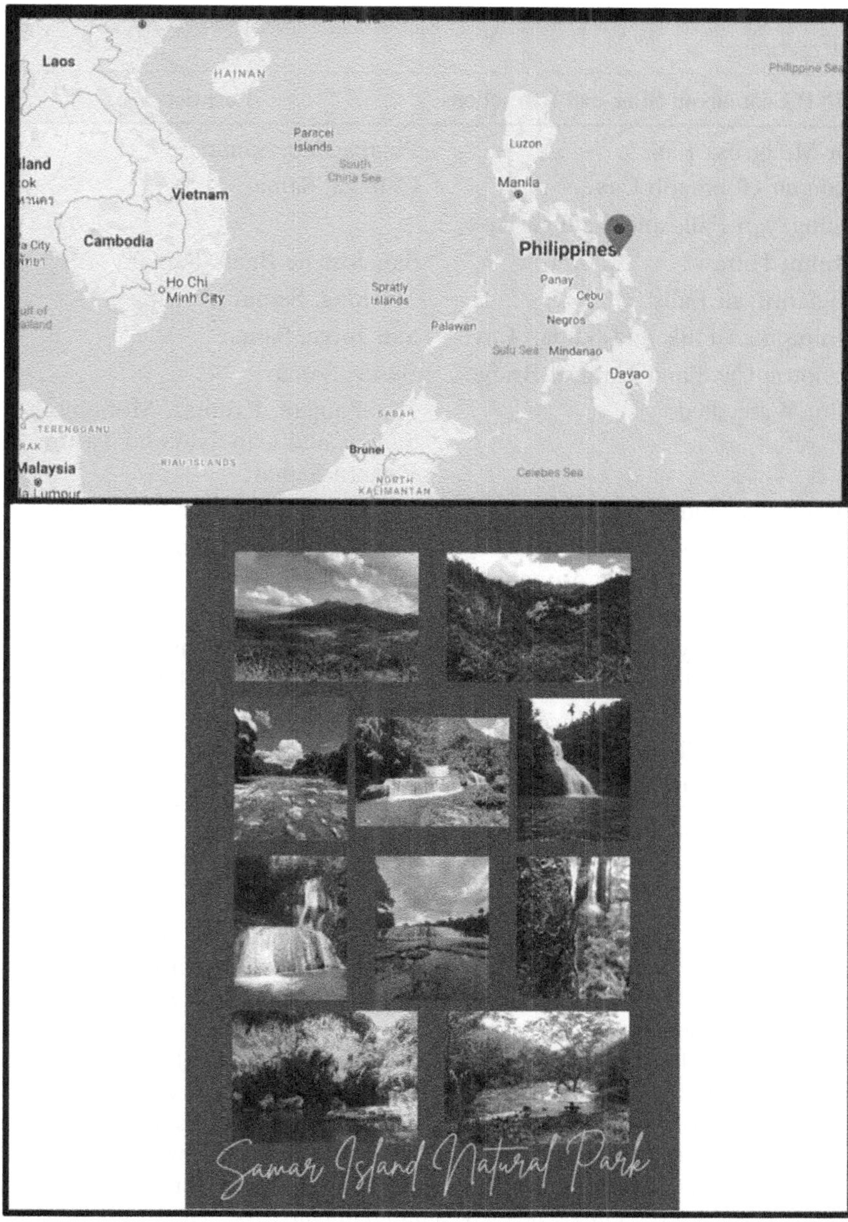

Fig. 11.1. Figural Representation of the SINP Ecotourism Sites and Attractions and Its Location in the Philippines. *Note:* Conceptualized by the researchers.

Table 11.1. Tabular Representation of the SINP Ecotourism Sites and Attractions.

SINP Ecotourism Sites and Attractions	Location
La Milagrosa Falls	Pinabacdao, Samar
Langun-Gobingob Cave	Calbiga, Samar
Lulugayan Falls and Calbiga River	
Mount Huraw	San Jose de Buan, Samar
Paglantuyan Falls	Motiong, Samar
Pinipisakan Falls and Sulpan Cave	San Jorge, Samar
Sohoton Cave and Natural Bridge	Basey, Samar
Ulot Watershed	Hinabangan, Paranas, Motiong in Samar, and Can-avid and Taft in Eastern Samar
Pinipisakan Falls	Las Navas, Northern Samar
Bulusao Watershed and Forest Reserve (Amandaraga, Amanhuray, Pange, and Banawan Falls)	Lawaan, Eastern Samar
Borongan, Maydolong, Balangkayan, Llorente Closed Canopy Forest	Borongan, Maydolong, Balangkayan, and Llorente, Eastern Samar
Taft Forest Philippine Eagle Sanctury (including Maharlika Falls)	Taft, Eastern Samar

Note: Conceptualized by the researchers.

hectares in the Greater Mindanao Faunal Region and the Eastern Visayas Bio-Geographic Zone, is a haven of biodiversity hosting numerous site-endemic species. Formed during the Pleistocene period, SINP's unique geological history enabled the colonization of various faunal species, fostering a rich diversity of wildlife. The park encompasses a variety of ecosystems, including lowland evergreen forests, transitional lower montane forests, mangroves, and more, with discontinuous forest corridors serving as vital watersheds and gene banks for the province. Given its representation of Greater Mindanao fauna, especially lowland forest taxa, SINP's conservation is paramount to safeguard critically endangered species like the Philippine Cockatoo, Philippine Eagle, and Philippine Crocodile. Its beauty lies in the harmonious blend of diverse landforms, forest types, and water features, while SINP's role as a microcosm of terrestrial and freshwater ecosystems makes it crucial for the preservation of threatened species, with numerous International Union for the Conservation of Nature (IUCN)-classified vulnerable, endangered, and critically endangered flora and fauna residing within its boundaries. Based on the first phase of the SINP-UNESCO Nomination Dossier, Criterion (vii) highlights SINP's intricate blend of land and water forms interwoven with diverse ecosystems, including lowland rainforests, montane forests, swamps, mangroves, limestone forests, and ultramafic soil forests, accentuated by mountain ranges, rivers, karst formations, waterfalls, and caves. Its allure lies in the complex, harmonious interplay of these features across six forest types, each offering a unique aesthetic experience throughout the year. Criterion (ix) showcases SINP as a microcosm of ecological and biological processes, hosting a wide array of terrestrial and freshwater ecosystems and plant and animal communities. It encompasses all known forest types, various headwaters, waterfalls, and river systems, serving as critical freshwater habitats for site, regional, and national endemic species. Criterion (x) on the other hand underscores SINP's significance in preserving biodiversity, especially threatened flora and fauna of outstanding universal value (OUV). The park shelters a remarkable number of IUCN-classified vulnerable, endangered, and critically endangered species, including 16 Dipterocarp species (*Hopea* and *Shorea* genera), three critically endangered and vulnerable reptilian species, five endangered mammalian species, and 15 avian species, all of which are Philippine- and Mindanao-endemic and classified as vulnerable, endangered, or critically endangered by IUCN.

The connection that local communities have with SINP is a manifestation of spiritual and recreational values, which is another integral aspect of its cultural and ecological significance. Indigenous practices and rituals often revolve around the park's natural features, and the values-based approach ensures that these spiritual connections are respected and maintained. Research findings reveal that locals primarily flock to attractions for wellness and recreational purposes, a manifestation of strong familial ties and affairs in their community. Moreover, the process promotes responsible recreational activities that allow local and foreign visitors to experience the park's beauty without compromising its values.

Economic sustainability through sustainable tourism is a crucial component of SINP's values-based approach. The government and other important stakeholder groups are expected to carefully plan and manage tourism activities that generate income for local communities while minimizing negative environmental and cultural impacts. The research findings and results show strong evidence that SINP attractions can significantly contribute to the local economy. Even though many attractions are still free of charge or accessible to the public, the volume of people or foot traffic in several attractions is a good indicator that this can be monetized. This economic value is essential for the long-term sustainability of both the park and its neighboring communities.

By implementing current tourism programs and projects, different stakeholders have experienced significant benefits and costs. In the context of its positive impacts, local communities, primarily residing within the SINP, generate income and experience profitability due to different livelihood and job opportunities. Some are related to *banig* weaving (local mat), food catering, assistance in various adventure-related activities, and serving as community tour guides. It also helps the government divert local communities' lifestyles and make them sustainable. Before tourism activities were implemented, the local communities concentrated on timber poaching or kaingin, which contributed to the destruction of the natural resources of SINP. In the context of education, the natural and cultural heritage and tourism of SINP are integrated with the curriculum, pedagogy, instruction, and competencies of different programs at the senior high school level, specifically on the Department of Education (DepEd) Basey and Calbiga Divisions in the province of Samar. Tour guiding and tourism promotions are some of the programs that are being offered. After the program, most students were assessed and certified by the Technical Education and Skills Development Authority (TESDA) regional office (Mercado et al., 2022).

SINP serves as a valuable resource for scientific research and environmental education. Hence, Scientific and Educational Values are an omnipresent endeavor on this island. It is, thus, evident as the primary study centers on its nomination for the next UNESCO World Heritage Site. Researchers from different backgrounds and disciplines frequently visit the island for research fieldwork and experiments. The values-based approach promotes using the park as a living laboratory. It encourages researchers and educators to study and share their findings about biodiversity, ecological processes, social sciences, and business. This adds to the park's significance and contributes to national importance and global scientific knowledge.

While SINP's values-based approach is commendable, it faces challenges in fully integrating economic benefits, social values, and politico-administrative affairs and climate. Balancing the needs of local communities, conservation, and tourism can be complex. Additionally, the approach's applicability to *living heritage* sites and other contexts warrants continuing examination. To address these challenges, SINP can benefit from employing social science research methods, utilizing digital technologies for monitoring and education, and engaging in community-led and coproduction ways and means. These tools can help refine and adapt the values-based approach to ensure that it continues to effectively preserve the unique significance of SINP.

Lastly, the government observed how local communities became proud of SINP's resources and their significance, which led them to promote and disseminate information to their community peers and visitors. It relates to the development of current studies associated with the positive impacts of tourism such as economic, sociocultural, and environmental aspects (Cellini & Cuccia, 2016; Patuelli et al., 2013; UNWTO, 2020). In conclusion, SINP's values-based approach sets a remarkable example for heritage preservation by recognizing the interconnectedness of cultural, ecological, and social values. By prioritizing these values, SINP safeguards its rich heritage and paves the way for a sustainable and harmonious future where nature and culture thrive together.

Samarnons and SINP's Issues, Constraints, and Opportunities: Rights-Based Approach

Voices of Samarnon Tourism Frontliners: The People's Organization

The efforts of the community are vital in pursuing their goals and objectives. People's organizations in the tourism and hospitality industry are nongovernment institutions that concentrate on achieving a specific goal, which benefits different stakeholders and uplifts their lifestyle and cultural environment. For SINP, the perspectives of the people's organizations are divided into five themes with different categorical identities. The five themes highlight an organized perspective of different people's organizations in SINP that concentrates on tourism and hospitality. The five themes are resources, stakeholders, opportunity, challenges, and stakeholders' initiative. Table 11.2 provides a brief overview of the five themes and their categorical identities.

Table 11.2. Tabular Representation of the Brief Description of the Five Themes and Its Categorical Identities.

Themes	Category	Brief Description
Resources	Physical and human geography Sites and attractions	It highlights the identity of SINP as a physical space in the island province and what can be visited by possible tourists.
Stakeholders	Government Nongovernment organization People's organization Responsible tourist Academe	It lists the associated primary and secondary stakeholders of SINP based on the perspectives of the people's organization.
Opportunity	Economic Environmental Pandemic-related	It determines the different positive opportunities that were engaged and experienced by the people's organization.
Challenges	Management Environmental conservation Hazard Sociocultural	It identifies the different problems encountered by the people's organization and other related stakeholders of SINP, especially in the development of tourism programs and activities.
Stakeholders Initiative	Operational Legislative Interpretation Personal Destination promotion Product development Support services	It highlights the programs, projects, and associated actions of the primary and secondary stakeholders on the development and conservation of SINP.

Note: Conceptualized by the researchers.

Under the resource theme, SINP is significant due to its available natural and cultural entities which are inculcated in the community and the environment. The resources are manifested in the physical and human geographical entities (focused on the areas of concern, location, climatic effect, and the overall scenery with its flora and fauna) and the available tourist sites and attractions that are developed through time.

Stakeholders' themes represent the different primary and secondary role players in the conservation and sustainable development of SINP. They provide a brief background of the key players and their specific duties and responsibilities. They also provide different programs and projects that would positively benefit them. SINP stakeholders are composed of the government sector (national, regional, and local), NGOs, people's organizations, tourists, and the academe. The government sector is responsible for the management and organization of SINP's programs and projects. This sector includes the DENR PAMB, DOT, Department of Public Works and Highways (DPWH, Department of Social Welfare and Development (DSWD), House of Representatives, Provincial, Municipal, and City Government institutions. The NGO contributes to facilities development, management, and initiatives that focus on the conservation of natural resources. ABS-CBN Foundation and the established Youth Organizations are some of the NGO support arms of SINP. The People's Organizations are responsible for the development and execution of their tourism products and services associated with an SINP site. These are local communities that allocate their time and effort to progressively sustain positive lifestyle growth through tourism and hospitality. Responsible tourists are visitors who contribute to the sustainability of SINP through proper waste disposal. It can involve international and national visitors who are motivated wholly or in part by interest in exploring SINP. Lastly, the academe is mandated to contribute to research and innovation initiatives in tourism and natural resource conservation.

The opportunity theme contributes to the experienced positive attributes from implementing sustainable tourism and hospitality, thereby benefitting different stakeholders. Economic, environmental, and pandemic-related opportunities are some of those attained by the stakeholders. The economic category provides sustainable contributions to changing the lifestyle of the stakeholders through income generation and alternative job opportunities. The environmental category, on the other hand, highlights the immense contribution of tourism and hospitality to sustaining the natural resources and the environment of SINP through direct and indirect initiatives and effects. Lastly, the pandemic-related category is focused on the benefits attained by the tourism and hospitality stakeholders, especially front liners, during the COVID-19 pandemic. It also highlights the retraining and re-evaluation of the process and practices of the tourism and hospitality stakeholders during the pandemic.

In the execution of programs and projects, challenges and problems may be encountered by the stakeholders. These are manifested through miscalculations during the implementation and also by external forces. In SINP's case, challenges are being faced on the management and organization, environmental conservation, external hazards, and sociocultural entities. Being a multi-geographical entity, the

SINP is governed by different governmental institutions. Thus, differences and inconsistencies in policies, standards, and guidelines are being experienced by the stakeholders. In terms of environmental conservation, SINP is rich in natural resources, specifically flora and fauna. Because of its richness and value, people committed crimes associated with illegal activities that led to the deterioration of SINP's natural scenery. Illegal logging and timber poaching are some of the common issues. Among the negative impacts of tourism are irregularities in carrying capacity which are also common issues. Natural and biological hazards are also significantly experienced by the stakeholders. It also highlights the issues arising from typhoons and the impact of COVID-19 pandemic. Lastly, sociocultural- related problems are also felt by the stakeholders, specifically insurgency, safety and security, health and sanitation, and moral and value-based problems.

From the problems encountered by the stakeholders, personal and communal initiatives may arise. The initiatives may contribute to the sustainability of SINP through responsible tourism and hospitality programs and projects. The initiatives are manifested through their tourism operations, legislative actions, interpretation, personal, destination promotions, product development, and associated support services. The operational initiatives concentrate on the normal processes and practices of different stakeholders in sustaining their tourism activities and how these affect other stakeholders. Legislative actions articulate the legal implications of the processes by providing appropriate policies and guidelines. Interpretation promotes the transmission of the significance and values of SINP through education and international recognition, such as the UNESCO World Heritage Sites. Personal initiative highlights the community's commitment and contribution to sustaining the environment and the local tourism industry. It also encourages other people to coordinate and collaborate for relevant and appropriate programs. Destination promotions are sustainable platforms on how to motivate potential tourists to explore and experience related tourism products and services in SINP. To continuously support the initiative of destination promotions, stakeholders are also exploring the possibility of product development based on the significance of the heritage property and their perspectives. Lastly, support services are also needed to complete the experience of potential tourists through infrastructure and superstructure development. Associated facilities and amenities on tourism, hospitality, and conservation are being implemented by the stakeholders.

SINP's Strengths, Weaknesses, Opportunities, and Threats (SWOT): Consultative Voices of the Samarnon

The SINP is endowed with several natural tourist attractions that provide it with several tourism advantages. It has the luxury to benefit from its nature-based tourism, specifically ecotourism, cultural tourism, adventure tourism, hard adventure, and educational tourism. It is generally blessed with undisturbed biodiversity. It is noteworthy that the conservation of natural areas, habitats, and wildlife is apparent. Both the public and private sectors are conscious of the need to preserve their rich biodiversity. Thus, the diversity and endemism of natural resources are higher than the average, which gives the island an overwhelming advantage over the rest of the Philippine destinations. These attractions are shared by three provinces: Samar, Northern Samar, and Eastern Samar. With this area, it can claim a good number of foreign and local tourist arrivals because of the several attractions present. Aside from the natural attractions, it also has several festivals that attract tourists to the different areas of the island. The festivals are celebrated in almost every district or town. Once the visitors decide to attend a festival, they are already confirmed visitors to the natural attractions that the district or town is offering. With this, tourism products in a district or town are being availed of by the tourists which in turn provide good economic activity for the locals. Another interesting tourist product is the presence of several delicacies distinct in a town or locality. The delicacies become instant favorites of the first-time visitors, and they usually bring the delicacies to their respective homes. The *Samarnons* (the local term for the community living in the area) are proud of their gastronomic heritage, and they are documenting their native delicacies so that this rich heritage is well preserved. In terms of accessibility, four airports and several seaports serve as gateways to the island, which offers a strategic route of trade through the Pan-Philippine Highway connecting Luzon to Mindanao. The presence of component cities that cater to a viable economy and other related industries creates an enabling environment for businesses. Noteworthy is the genuine involvement of all the stakeholders, as the SINP is collaboratively managed by the DENR, DOT, the Provincial Governments and their respective LGUs, NGOs, POs, and the Church. The rising popularity and influence of NGOs and POs also positively affect SINP. As early as 2005, several NGOs and POs have been actively involved in the conservation and management of SINP. A good example is the Province of Samar which has an established souvenir shop which supports local trade and crafts. The presence of Intact Community Solidarity and Visible Community Involvement and Participation is clearly evident, resulting in tourism acceptance in the social fabric of communities. A concrete example is the conversion of locals from illegal loggers and insurgents to tourism movers. The illegal loggers became protectors of the environment because they knew the advantages of tourism in their lives. The source of their tourism products is the natural beauty of their place; therefore, they protect and preserve their natural environment. They are able to do this because of the strong, solid community and involvement. The LGUs have generally good local and provincial tourism policies, planning, and development movements. Most of the municipalities or cities have a Tourism Code of their own. Local leaders exhibit political will through programs or projects on the environment and culture or heritage that are aligned with both provincial and local governments. With the COVID-19 pandemic, stakeholders created Guidelines and Protocols for Tourism and Hospitality Operations. In terms of marketing, SINP is supportive and interested in promoting and marketing their respective destinations to tourists. The DOT is giving its all-out support for all the marketing efforts of SINP. For its human resources, *barangay* officials are very supportive of and receptive to the idea that they will take part in the tourism development of their community. The POs and their members are the tourism front liners, providing sustainable tourism products and services to tourists. As for the financial aspect, SINP, as a protected area, is given a considerable budget for its operations and development.

However, there are several challenges that need immediate action through a multi-sectoral approach. The SINP Tourism-based Organization and POs for some of the sites are nonexistent which is crucial to the management aspect. Aside from this, there are problems with the gathering and monitoring of tourism statistics which are vital for the decision-making of LGUs. Neglecting the collection of crucial data will further exacerbate the presence of apathy in the government sector especially in planning, implementing, and monitoring of projects and programs. Furthermore, there is inconsistency in positions and functions which causes confusion. For instance, the Municipal Tourism Office is under the jurisdiction of another department of an LGU. A solid and stable organizational structure in terms of tourism governance must be instituted in all LGUs. The implementation of the Carrying Capacity for SINP is also questionable in terms of its consistency. Thus, there are problems with the maintenance and development of tourism facilities and their environment. There are destinations that accept all visitors who are willing to stay without taking into account the carrying capacity. On the other hand, ownership and management issues for some of the SINP sites are still major concerns. For instance, ownership issues exist between the municipalities of Motiong and Jiabong over the control of Paglantuyan Falls. This will further lead to problems at accessibility points of several attractions such as the Borongan-Maydolong-Balangkayan-Llorente Closed Canopy Forest, which is quite confusing for visitors. There are also inconsistencies in the expansion of benefits of tourism to other stakeholders. This leads to problems in the distribution of economic benefits to locals or to the grassroots. Because of the pandemic, there is also an apparent decrease in average occupancy rates for the accommodation facilities around the island. Generally, the quality of life of the locals is apparently affected. Problems with solid and liquid waste management are also big challenges for the government. Intergovernmental and multi-sectoral approaches are needed to mitigate these problems to prevent further ramifications. Among the problems that need similar interventions are insurgencies and terrorism because of the presence of rebel groups around the island (Mercado et al., 2022). Aside from these, issues of political killings are apparent which have already affected the image of the entire island.

Among the SINP's opportunities was its legal strengthening through Presidential Proclamation No. 744, s. 1996 or "Declaring the Rainforests of the Provinces of Northern Samar, Eastern Samar, and Western Samar as Forest Reserve" (Malacañang, 1996). In addition, there are more than 10 legislations, issuances, and proclamations, as well as local ordinances that are aimed toward sustainable tourism development. With these legal issuances, SINP is assured of its protection, preservation, and conservation. With its rich biodiversity and natural attractions, SINP is poised to be part of the UNESCO World Heritage Sites, which will open several opportunities to the whole island. There are several familial practices for the conservation and preservation of natural resources that are transmitted naturally which could benefit the tourism industry in particular.

In terms of the area's climate, the SINP is endowed with a tropical rainforest climate, which is warm, humid, and rainy all year around. With this kind of climate, international tourists who wish to escape from the cold winters in their countries would easily choose it as their destination. Local and international publications recognize SINP as one of the most important areas in terms of biodiversity in the island of Samar and within the Greater Mindanao Faunal Region. The island has major ore commodities such as hematite and magnetite chromate, nickel, and manganese. Meanwhile, copper, minor zinc, and lead are usually in the primary form of sulfide and associated with abundant pyrite. There are also extensive caves in the island's limestone karst, harboring a unique biota that remains largely unexplored (Hunnam et al., 2004). On the economic side, the three provinces that comprise SINP are basically driven by agriculture, fishery, and forestry industries. Trading is on top of all other commercial activities on the island followed by the service sector which includes restaurants, lodging houses, hotels, and resorts. Although small in number, this part of the service sector is available. While the COVID-19 temporarily shut down different tourism and hospitality establishments in the country, the industry is taking actions and programs to minimize the spread of the virus and to strictly implement the mandates and protocols conceptualized by the Inter-Agency Task Force for the Management of Emerging Infectious Disease (IATF-MEID) of the Philippines (DOT, 2020).

Due to the significance of the natural environment of SINP, there are numerous species of flora and fauna that are endemic and endangered. This prompted the community to engage in illegal livelihood opportunities such as logging and poaching. Through the forest resources, people could earn more compared with their normal livelihood opportunities. But the illegal activities have resulted in the degradation and destruction of natural resources. Another negative factor is unsustainable infrastructure development attributed to the SINP. Development is always needed to progressively provide basic goods and services to different stakeholders. Part of societal development is infrastructure, such as roads, bridges, transportation hubs, and superstructures. Construction of some infrastructures deteriorated and degraded the natural and cultural heritage significance of SINP. Based on the records of the Philippine Statistics Authority (PSA), Eastern Visayas is one of the poorest regions in the country. It renders a negative image for the region, specifically on the availability of resources to fulfill basic needs and wants, infrastructures and related facilities, and support services for local communities and visitors. It also reflects the problems with sustainable education and lifelong learning among local communities. Records show that there are concerns about the rate of enrollment and graduation of locals from basic to secondary education. Also, basic healthcare facilities such as private and public hospitals, clinics, and barangay health centers, and their related human resources are still a problem in the region. Unfortunately, the region is also prone to different natural and biological hazards. It could affect the lifestyle of the community given that the residents can experience meteorological (typhoons, flash floods, storm surges), geological or geomorphological (seismic and mass movement), human-induced (fire, pollution, violence, disease such as COVID-19, human–wildlife conflict, illegal activities, and warfare through insurgency) factors, and climate change. Because of the hazards present, for instance, the COVID-19 pandemic, limited interactions are implemented to diminish the risk of transmission and minimize the number of positive cases. Due to this problem, tourism activities and associated actions are limited resulting in the decreasing positive result and impact to the stakeholders in the economic, socio-cultural, and environmental aspects. Another problem is insurgency and safety and security, specifically on political killings. The presence of Communist Party of the Philippines - National Democratic Front Terrorists (CNTs) is one of the major concerns of the local and national law enforcers. Policies and management are also considered significant problems. Inconsistencies on the initiatives and policies provided by different LGUs concerned are one of the problems highlighted. Although there are good practices implemented by some LGUs, it is significant if common approaches and strategies are to be implemented since it is a multi-geographical entity. Likewise, problems in the implementation, monitoring, and evaluation of local policies and guidelines were also observed. For instance, evidence proves that on waste management, there are littering cases observed for most SINP sites and attractions even though there are policies and regulations being mandated in the area. It is associated with problems related to self-discipline of the local communities and visitors. They have ideas on the policies mandated by the stakeholders, but the concept of "ownership" from the local community and the impact of the "value of tourist experience" are being practiced.

Understanding the connection between the internal and external environments provided meaningful recommendations on programs and projects to the local community and potential visitors for tourism. The presence of SINP's SWOT highlights sustainable platforms that are naturally experienced by stakeholders. The connection between the strengths and the opportunities provides helpful ideas and concepts, the significance of strengths to minimize the impacts of threats, the significance of opportunities to minimize the impacts of weaknesses, and the expected harmful impacts of weaknesses and threats. Ultimately, the four areas provide meaningful approaches to resolving problems and constraints through better understanding the rights of the stakeholders and of the rights-based approach.

SINP Sustainable Tourism and Strategic Development Framework

Through the integration of the SINP's sense of value identity and perspectives of primary and secondary stakeholders, a 10-year sustainable tourism strategic development framework was conceptualized. Its mission statement "institutes the foundation of ethical policies and promotes multi-stakeholder involvement that would result in sustainable impacts to the economic, socio-cultural, environmental, and politico-administrative aspects." And through the implementation of the strategic framework, the stakeholders envisioned that "By 2031, SINP will be a Sustainable Tourism Destination concerning its economic, socio-cultural, environmental, and politico-administrative aspects guided by Ethical Policies formulated through Participatory Governance" (Provincial Governments of Samar, Northern Samar, and Eastern Samar, DOT, DENR, UST GS-CCCPET & CTHM, 2022, p. 146). The development of SINP's tourism products and services must be collaboratively planned and implemented by primary and secondary stakeholders, including the National and Local Government Units, Regional Security Forces, People's Organizations, Tourism Business Sectors, Local Communities, and Tourists. Fig. 11.2 presents the SINP's Sustainable Tourism and Strategic Development Framework.

In the fulfillment of the SINP vision and mission and harnessing the interconnectivity of the value and rights-based approach, the stakeholders were concentrated on the implementation of four strategic directions, namely *HEAD* or the Holistic Economic and Accessible Development (Bottom-up Local Economic Development), *Kabilin* or the Protection and Safeguarding of SINP's Socio-cultural Values and Significances, Full Environmental Security, and *LEAPP* or Leadership Excellence through Active Participatory Partnerships.

A holistic approach to development is crucial in determining the sustainability of communities in both urban and rural settings. Specifically, in rural areas like Samar Island, holistic development is vital in ensuring that the citizenry feels government services, including those living in far-flung areas and are seldom reached by public service delivery. Rural areas have been the primary settlement of about half of the Philippine population (IFAD, 2004). According to Gabito (2012) and Aya-Ay et al. (2020), those living in rural areas are the indigents or people with low incomes, specifically the landless farmers, fisher folks, people in upland areas, and Indigenous people (IP) who depend on agricultural activities for a living. Nevertheless, these rural communities' unique settings, outstanding features, and natural resources theoretically have the potential and can be harnessed for tourism purposes, thereby generating jobs and alternative livelihood. The *HEAD:* Holistic Economic and Accessible Development, focusing on a Bottom-up Local Economic Development approach, serves as a strategic direction that will capitalize on SINP's grandeur and maximize its potentials to be a self-sustaining economic enterprise with high regard for the people and the environment.

"Kabilin" is a Waray term for "heritage." It refers to the significance of cultural values and traditions of the community living within SINP in developing their everyday lifestyle. The second strategic direction focuses on protecting and safeguarding SINP's sociocultural values and significance. It is aligned with the United Nations (UN) Sustainable Development Goals (SDGs) (UNWTO, 2023) on "sustainable cities and communities (#12) and partnerships for the goals (#17)" and the UNWTO Global Code of Ethics for Tourism (1999), specifically on the "factor of sustainable development (3), user of the cultural heritage of mankind and contributor to its enhancement (4), and a beneficial activity for host countries and communities (5)." The strategic direction provides five programs for implementing platforms and activities. These programs are sustainable developmental frameworks and application-based plans for sociocultural aspects, the

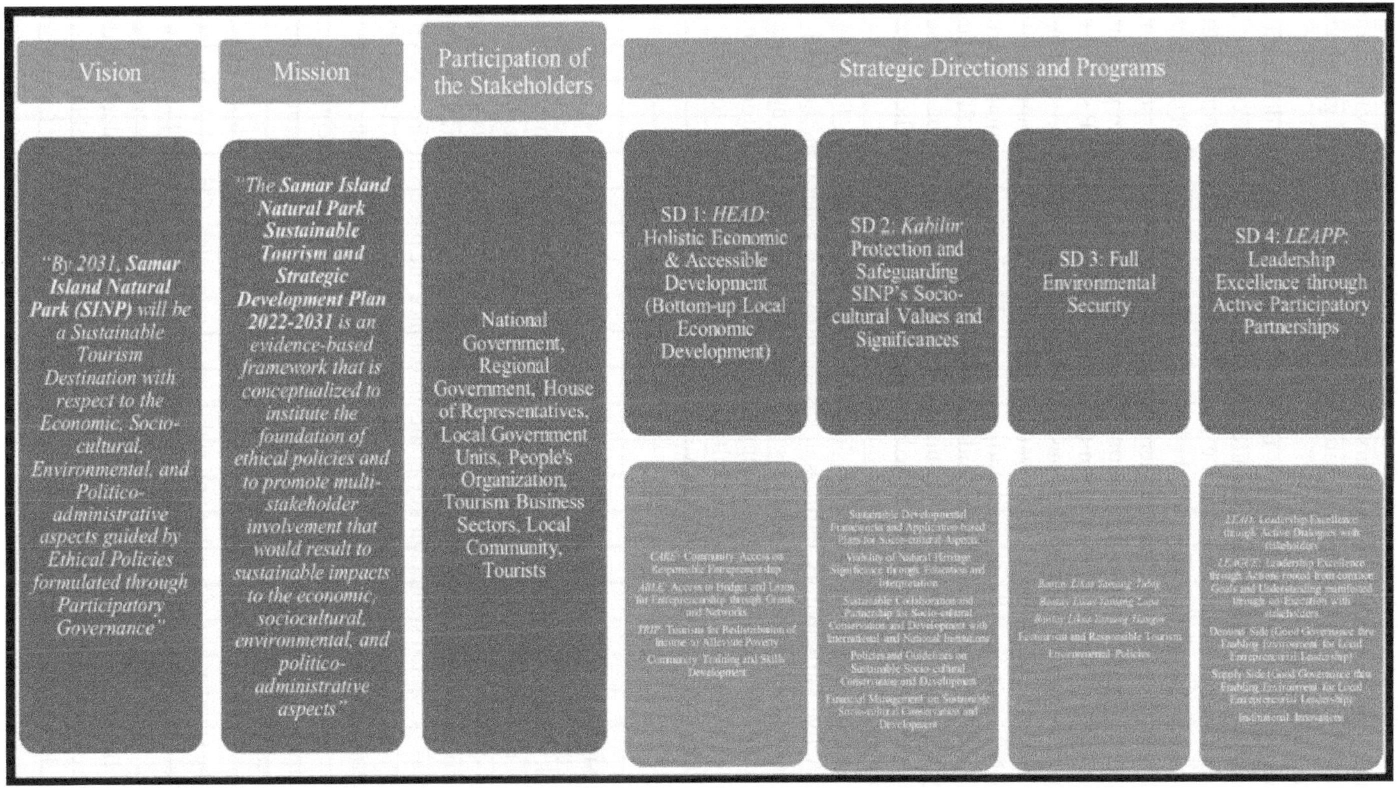

Fig. 11.2. SINP's Sustainable Tourism and Strategic Development Plan 2022–2031. *Source:* Provincial Governments of Samar, Northern Samar, and Eastern Samar, DOT, DENR, UST GS-CCCPET and CTHM (2022).

viability of natural heritage significance through education and interpretation, sustainable collaboration and partnership for sociocultural conservation and development with international and national institutions, policies and guidelines, and financial management.

The natural environment of SINP is the most crucial element of its declaration as a natural park. Its rich natural resources have to be protected and preserved. It is of paramount consideration that the whole island will embrace sustainability through which the natural park will be able to secure its resources. This entails great efforts from all the stakeholders to ensure that SINP will sustain its resources from generation to generation. There is a need for concrete programs and policies to attain total environmental security. Each area of SINP should be given much consideration regarding sustainability. All its landforms, the things that are beneath and above it, all the water systems, everything that comprises the whole SINP must be fully secured and sustained. Total environmental security will keep the SINP free from abuse and damage. All stakeholders of SINP must seriously consider committing to genuine sustainability which can be achieved through good programs that are geared toward the preservation of the environment and ensuring its sustainability.

The overall goal of the strategic direction *LEAPP:* Leadership Excellence through Active Participatory Partnerships is for it to be a general method rather than a specific tool. This approach aims to facilitate the inclusion and engagement of individuals or groups in the planning, designing, executing, monitoring, and evaluating of policies through consultative or participatory ways to achieve accountability, transparency, and active citizenship.

Conclusion

This book chapter is a captivating journey into the heart of SINP, an untouched sanctuary nestled on the easternmost side of the Philippines. By immersing in the rich landscapes and biodiversity of SINP, it became evident that this ecological frontier holds immense potential for sustainable tourism development. This book chapter has presented significant key points to balance the preservation of SINP's ecological integrity with harnessing its tourism potential, aiming to benefit the local communities while celebrating the unique facets that make SINP a distinctive destination. The study's primary objective was to conceptualize an STSDP using value and rights-based approaches, aligning with the requirements for a potential UNESCO World Natural Heritage Site application.

The discourse extends to the lived experiences of local communities, underscoring the significance of a value and rights-based approach. Critical discussions on natural and cultural heritage significance and the SINP STSDP Framework provide a comprehensive foundation for stakeholders to navigate toward its common objectives. Through the integration of SINP's intrinsic value and the perspectives of primary and secondary stakeholders, a 10-year sustainable tourism strategic development framework was conscientiously crafted. This framework aspires to usher in sound policies and foster multi-stakeholder involvement, ultimately leading to sustainable impacts on the economic, sociocultural, environmental, and politico-administrative dimensions. The stakeholders envision a future where, by 2031, SINP will become a model of sustainable tourism, guided by environmental and human-centric policies formulated through participatory governance. The strategic directions are a guiding methodology that empowers individuals and groups to actively participate in the development process. This inclusive and consultative approach ensures accountability, transparency, and active citizenship.

Ultimately, the STSDP 2022–2031 presented in this chapter represents a significant step forward in achieving sustainable development in the context of the fastest-growing industry in the world today: tourism. It envisions delivering economic benefits to Samar Island communities while safeguarding its ecological and sociocultural identity. This comprehensive and holistic plan provides a blueprint for a harmonious future where nature and communities thrive together, setting the stage for a more sustainable tomorrow for Samar Island. By implementing value and rights-based approaches to tourism development, customization can be executed in a sustainable manner manifested on its products and services – emphasizing the significance of place as basis for developing programs and projects and gathering stakeholders' perspectives on both the positive and negative implications.

Acknowledgments

The book chapter is based on the Samar Island Natural Park (SINP) Sustainable Tourism and Strategic Development Plan 2022–2031. It is part of the joint project of the Provincial Governments of Samar, Northern Samar, and Eastern Samar, the Department of Tourism Region VIII, the Department of Environment and Natural Resources Protected Area Management Board Region VIII, the University of Santo Tomas Graduate School – Center for Conservation of Cultural Property and Environment in the Tropics, College of Tourism and Hospitality Management, and the UST Sentro Turismo for the UNESCO World Heritage Nomination of SINP.

References

Australia ICOMOS. (2013). *The Burra Charter: The Australia ICOMOS Charter for places of cultural significance*. Australia International Council on Monuments and Sites. https://australia.icomos.org/wp-content/uploads/The-Burra-Charter-2013-Adopted-31.10.2013.pdf. Accessed on September 09, 2023.

Avrami, E., Mason, R., MacDonald, S., & Myers, D. (2020). *Values in heritage management: Emerging approaches and research directions*. Getty Conservation Institute.

Aya-Ay, A. J., Gabriel, K. J., Guilatco, D. P., Maglaya, Y. S., Mendoza, M. R., Tacneng, M. A., & Andalecio, A. B. P. (2020). Pro-poor tourism concept in the context of selected destinations in Ilocos region, Philippines. *Journal of Tourism, Hospitality & Culinary Arts*, *12*(1), 69–91. https://fhtm.uitm.edu.my/images/jthca/Vol12Issue1/Chap_5.pdf

Cellini, R., & Cuccia, T. (2016). UNESCO sites as public goods: Comparative experiences in Italy. *Revista de Economia Contemporânea*, *20*(3), 553–569. https://doi.org/10.1590/198055272037

DOT. (2020). *Health and safety manual governing the operation of tourism sites and activities under the new normal in Region VIII* [Unpublished document]. Department of Tourism Region VIII.

Gabito, O. A. (2012). *Growing rural employment through community-based tourism*. Working Paper 9. https://ils.dole.gov.ph/growing-rural-employment-through-community-based-tourism/. Accessed on September 05, 2023.

Hunnam, P., Ong, P., & Talisayon, S. (2004). *Samar Island biodiversity project. Mid-Term Evaluation Report - PHI/99/G31 SIBP*. United Nations Development Programme.

ICOMOS. (2019). *Water heritage – Questionnaire on development of ISC, mission statement and statement of significance for water as cultural heritage*. International Council on Monuments and Sites. https://www.icomos.org/en/member-area/64180-water-heritage-questionnaire-on-development-of-international-scientific-committee-mission-statement-and-statement-of-significance-for-water-as-cultural-heritage. Accessed on January 20, 2024.

IFAD. (2004). *Philippines*. International Fund for Agricultural Development. https://www.ifad.org/en/web/operations/w/country/philippines. Accessed on September 05, 2023.

Malacañang. (1996). Declaring the rainforests of the provinces of Northern Samar, Eastern Samar, and Western Samar as Forest Reserve. Malacañang.

Mason, R. (2008). Be interested and beware: Joining economic valuation and heritage conservation. *International Journal of Heritage Studies*, *14*, 303–318. https://doi.org/10.1080/13527250802155810

McClelland, A., Peel, D., Hayes, C.-M., & Montgomery, I. (2013). A value-based approach to heritage planning: Raising awareness of the dark side of destruction and conservation. *Town Planning Review*, *84*(5). https://doi.org/10.3828/tpr.2013.31

Mercado, J. M. T., Andalecio, A. B. P., Granado, G. G. G., & Zerrudo, E. B. (2022). Pagbalangkas: Issues and opportunities on natural heritage sustainability of Samar Island Natural Park (SINP), Philippines – Basis for inclusive organizational management framework. *Journal of Tourism, Culinary, and Entrepreneurship*, *2*(2), 226–250. https://doi.org/10.37715/jtce.v2i2.2728

Patuelli, R., Mussoni, M., & Candela, G. (2013). The effects of world heritage sites on domestic tourism: A spatial interaction model for Italy. *Journal of Geographical Systems*, *15*, 369–402. https://doi.org/10.1007/s10109-013-0184-5

Provincial Governments of Samar, Northern Samar and Eastern Samar, DOT Region VIIIDENR – PAMB Region VIIIUST GS-CCCPETUST-CTHM. (2022). *Samar Island Natural Park (SINP) sustainable tourism and strategic development plan 2022–2031* [Unpublished document]. Provincial Governments of Samar, Northern Samar and Eastern Samar, Department of Tourism Region VIII, Department of Environment and Natural Resources – Protected Areas and Management Bureau Region VIII, University of Santo Tomas.

UNWTO. (1999). *Global code of ethics for tourism*. United Nations World Tourism Organization. https://www.unwto.org/global-code-of-ethics-for-tourism. Accessed on September 04, 2023.

UNWTO. (2020). *Sustainable development*. United Nations World Tourism Organization. https://www.unwto.org/sustainable-development. Accessed on October 01, 2023.

UNWTO. (2023). *Tourism & sustainable development goals*. United Nations World Tourism Organization. https://tourism4sdgs.org/tourism-for-sdgs/tourism-and-sdgs/. Accessed on September 04, 2023.

Chapter 12

Ecotourism Destinations in Nueva Ecija: Concepts and Practices Toward Tourism Marketing and Sustainability

Celyrah B. Castillo and Marie Jel D. Bautista

Central Luzon State University, Philippines

Abstract

In the realm of postpandemic perspectives, tourism was awakened to see that atop economic viability is resiliency. Shift in the demand to visit ecotourism destinations became a valuable opportunity to the local community to capitalize on natural resources as primary precursor for livelihood, substantiating their communal role as bailiffs in the conservation of protected landscapes and mitigators of the negative tourism implications. As tourism initiates in rethinking sustainability and inclusivity, the postpandemic paradigm in destination planning became anew. Aside from being an agritourism destination, Nueva Ecija has been promoted as an emerging ecotourism destination. Various literature has evaluated local ecotourism destinations in the Philippines, yet few literature examined the potential of Nueva Ecija. The emerging popularity of the province's ecotourism postpandemic beckons the inflow of tourists compelling for an in-place policy toward sustainability. The study aims to underscore the destination branding of Nueva Ecija as a tourist destination using a strategic framework derived from ecotourism concepts and best practices. Qualitative data from various tourism stakeholders were collected, and internal and external factors were analyzed. Results of the study may serve as a foundation toward a unified framework to assess and strengthen ecotourism destinations toward sustainability and tourism destination resiliency.

Keywords: Destination branding; ecotourism; destination image; SWOC analysis; sustainability

Revisiting Sustainable Tourism in the Philippines, 219–236

doi:10.1108/978-1-83753-678-820241012

Rationale

The COVID-19 pandemic adversely affected the global economy, disturbed notably the tourism and hospitality industries. Due to the restrictions that subjected most of the areas to lockdown to reduce the spread of disease, the aftermath was a huge decline in tourism-related businesses. The downturn on domestic destinations took place for more than a year yet awakened the masses to embrace a new destination preference – several local ecotourism destinations emerged as the risks abated. As local destinations begin to recover, a comprehensive risk assessment and management enfold postpandemic phase on various ecotourist destinations. The study explored the Nueva Ecija ecotourism destinations which emerged after the pandemic and assessed sustainable practices and existing tourism products and services in reference to the strengths, weaknesses, opportunities, and challenges (SWOC) analysis. With this, feasible action plans in the postpandemic perspective have been proposed to address challenges in bringing forward unique propositions of ecotourism products and services. To objectively regulate proper visitor management among ecotourism destinations, allocation of carrying capacity has been substantially recognized to prevent environmental degradation alongside gravitating the magnitude of an accordingly administered sustainable ecotourism in the province. Thus, findings of the study can serve as a basis for a standardized strategy to manage sustainable and resilient ecotourism destinations properly and consistently in Nueva Ecija.

Ecotourism in Nueva Ecija

Tourism industry experienced numerous challenges with regards to the adherence to environmental paragons aligned with the marketability of ecotourism, including profit generation, of which outturns economic advantages yet at the outlay of environment (Choi et al., 2011). In ecotourism, being able to enjoy nature is not the sole element to consider but also its impacts to societal and economic gains, and the state of equilibrium among the tourism pillars.

Nueva Ecija is known as the biggest province and rice producer of Central Luzon due to its vast agricultural land. Nueva Ecija, an agricultural province, famed not only as the rice granary of the Philippines but also the prime producer of onions across Southeast Asia next to Vietnam and Indonesia. Aside from the agricultural facets, the province also possesses promising tourist spots and activities which attract more potential visitors, especially those who prefer nature sightseeing and activities. According to the tourism statistics as of 2022 of the Department of Tourism (DOT), the province placed sixth in terms of tourist arrival as compared with other provinces in Central Luzon. The province of Nueva Ecija was chosen as the locality of the study because of the various emerging ecotourism destinations offering natural pristine aesthetics of the province and nature-themed activities and facilities. The provincial government of Nueva Ecija started promoting the province as an ecotourism destination prior to the onset of pandemic until it paved to further market the ecotourism destinations in the postpandemic because of the emerging travel preferences to nature and ranked "excellent" based on visitors' evaluation of Nueva Ecija as a tourism destination based on historical value, uniqueness, and natural aesthetics (Pagoso & Katigbak, 2009; Viernes & Aquino, 2020; Vigilia, 2020).

Pillars of Ecotourism Vis-à-vis Postpandemic Travel Preference

Sustainable tourism as described by the United Nations World Tourism Organization (UNWTO) is a form of tourism with reference to addressing the current and future implications across the pillars of environment, social, and economy, anchored in delivering the needs and demands of various stakeholders such as tourists, service providers and destination managers, industry, and host community (Desai, 2022). As a subset of this concept, ecotourism is induced with responsible travel while improving the welfare of local people (Roxas et al., 2020). Ecotourism is a sustainable form of natural resource-based tourism. It focuses primarily on experiencing and learning about nature, its landscape, flora, fauna, and their habitats, as well as cultural artifacts from the locality (Fung & Wong, 2007). Whereas sustainability for ecotourism destinations holdings viable advantage to survive in today's era and saturated tourism market, taking note of innovative practices that are essential to sustain long-term benefits and of a collaborative effort between the service providers and stakeholders (Fernando, 2015; Shariffuddin et al., 2020).

Philippines, as tourist destination preference, among its most established present type of tourism, was known for social, cultural, recreational, and nature tourism, whereas destination preference still varies among classifications and with the growing concern of safety, most of these preferences motivate domestic travel (Chin et al., 2021). The COVID-19 pandemic resulted in the closing of several business enterprises and establishments in the tourism and hospitality industry and raised a huge percentage of unemployment and jeopardized fear associated with health and safety (Aruta & Guinto, 2022; Chin et al., 2021; Roxas et al., 2020). Upon the lifting of restrictions, domestic tourism discovered a significant role in jump-starting tourism. This paved a new route in rethinking Philippine tourism toward recovery and in promoting domestic travel through adopting new channels and in rediscovering destination preferences in the post-COVID-19 era (Bautista, n.d.). Tourists began to embrace domestic travel to nature-based destinations because of new travel pattern, motivation, and intention (Barr et al., 2021; Ivanova et al., 2021; Kusumaningrum & Wachyuni, 2020; Mayo et al., 2021; Robinson, 2020; Yoo et al., 2022).

Concepts and Best Practices of Ecotourism in the Philippines

Ecotourism corresponds to the natural environment and relevant tourism activities; however, ecotourism arises distinct characteristics because of the concepts of sustainability which crafts equilibrium among all tourism pillars. Ecotourism, in its fact-based construct, is a manifestation of natural travel that reinforces safeguarding and biodiversity, ameliorates the entity of local community, assists their educational initiatives, and incorporates the involvement of tourists through a form of education in relation with responsible tourism (Bansil et al., 2015;

Bautista, 2021). In Ocampo et al. (2018), the Philippines as a developing country has yet to capture the efficacy of ecotourism activities unless supported by the local community and the government.

Ecotourism holds substantial apprehension through environmental preservation and conservation, however noticed to comprise damaging effects on the environment, society and community.

> In its every scope, ecotourism is a motivator of change. It shapes alliances between people and the environment and among other cultures and ways of life. Tourists are in search of new experiences away from their routine through traveling to ecotourism areas. Residents living near ecotourism sites perceive an enhanced standard of living and life opportunities parallel to the development of potential destinations, and environmentalists together with the government and other concerned sectors yearn for the protection and conservation of ecology.
>
> (Bautista, 2021, p. 18)

Relevant with the increasing number of tourism establishments because of economic gains in ecotourism, sustainability among tourism establishments has been questioned since most were focused on the overwhelming effect of tourism receipts, while stability and resilience have been blurred by the milieu of economic expansion (Beall et al., 2021; Padin, 2016; Rivera & Gutierrez, 2018).

Despite the aesthetic variances, existing and emerging ecotourism destinations must capture its truest context as a platform in uplifting community and stakeholders, as a framework for conservation, as a scaffold for social and cultural heritage preservation, as a promoter of homegrown produce and local gastronomy, as a pillar for sustainability, as a component for global development, as a form of education to tourists, and as a motivator of change (Bautista, 2021; Cobbinah, 2015; Gutierrez, 2021; Pestek & Nikolic, 2011; Ziegler et al., 2021).

Destination Branding of Nueva Ecija as an Ecotourism Destination

The growth of tourism in the Philippines is largely driven by visits to natural attractions (Rivera & Gutierrez, 2018). Ecotourism acts as a valuable source in the uplift of tourism in the country, making it more prone to exploitation and deterioration of natural aesthetics and host environment including the community residing in the destination. According to the DOT (2011), paramount of their goals is to develop Filipino destinations and products. This is relevant with the concept of destination branding wherein a tourist site markets itself using a particular theme or concept that exhibits homegrown products of the location. The notion of destination branding and on how it directly impacts visitor influx has been explored by various studies which argued that the tourists' demand for experience economy served as a roadmap for destinations to market themselves as unique brand (Alejandria-Gonzalez, 2016).

The challenge emerged on how a local destination would develop and promote corresponding mechanisms that would concede to offer varied tourism experiences for visitors. Relevant with this emerging nature-based destination preference, the flagship of ecotourism destinations in Nueva Ecija is the *Minalungao* National Park (MNP), a protected landscape under the initial component of Republic Act 7586 also known as the National Integrated Protected Areas System (NIPAS) Act, and one of the few remaining natural environments in the region north of Manila (Pader et al., 2017). Since the province has been keen in the development and promotion of MNP, the other ecotourism destinations are left underdeveloped. As result of this, the tourism orientation of service providers to produce local products and service is also affected. In turn, the tourism orientation of visitors is also influenced by the promotional and marketing strategies of service providers and local government. Thus, this served as a context to address destination branding through the assessment of the perceptions of tourism service providers and tourists as associated with destination branding which could be merged to attain commonality in identifying the tourism destination branding of the province of Nueva Ecija based on the constructs of destination image, destination quality, destination awareness, destination personality, and destination culture (Tsaur et al., 2016).

Destination branding can assist tourists in furthering their perceptiveness of the destination as an evaluation of tourism experience (Tsaur et al., 2016). Similarly, destination branding consists of the demand-side image and the supply-side image assumed by tourists (Pike et al., 2010; Tsaur et al., 2016). To support this, there are six benefits of tourism destination branding: (1) reducing the choice, (2) reducing the impact of intangibility, (3) conveying consistency across multiple outlets and through time, (4) reducing the risk factors associated with holiday decision-making, (5) facilitating precise segmentation, and (6) providing a focus for the integration of producer effort (Clark et al., 2000; Tsaur et al., 2016). These support the definition and context of destination branding; destination service providers and managers must project a destination brand identity using the destination's unique characteristics and activities as means to be differentiated and distinguished among competitors. To support this, service providers and marketers may explore both sides of supply and demand using the strategic brand analysis framework in Konecnik and Go (2008). With focus on the supply-side aspect of destination branding, a brand identity model was proposed consisting of these elements: vision, mission, values, benefits, personality, and distinctive preferences (Konecnik & De Chernatony, 2012).

In Tsaur et al. (2016), destination branding dimensions have been deduced as: (1) *destination image* refers to the attitudinal concept as result of beliefs, impressions, and ideas of tourists in a destination; (2) *destination quality* is based on the perceptions of tourists with the destination's ability to meet their expectations; (3) *destination awareness* refers to the knowledge and background of tourists with the tourism destination; (4) *destination personality* are the set of human qualities linked with the tourism destination; and (5) *destination culture* are the cultural aspects and characteristics of a tourism destination. Nueva Ecija's destination branding concept could progress in addressing the factual and genuine depiction of the province considering the components of image, quality, personality, awareness, and culture. The construct of branding has been pictured with a wider array of opportunities for the province to maximize its potential as an ecotourism destination. With this, the province has been perceived beyond its natural aesthetics but highlighted the significance of stakeholders and respective perceptions and the inclusion of the local community who highly depends on the destination.

Proper market segmentation could help augment the destinations to align Nueva Ecija's products and services with the market profile of diverse market characteristics and future needs. The presence of night life has been viewed as less essential in considering the destination image of the province. Most stakeholders considered relaxation and nature activities to be the selling point of Nueva Ecija.

Tourism destination quality is a branding component based on continuous quality improvements which benefits the destination. In Seakhoa-King et al. (2020), destination quality is mentioned unparallel with the definition of the word "value"; associating destination quality with the term "value" can provide misleading impressions. At the destination level, the tourism product comprises a bundle of components including accommodation, food, and activities; while the overall evaluation of the tourists includes services, products, and experiences.

Accommodation has been agreed to contribute to the destination quality of the province in view of how the stakeholder acknowledged distinctiveness of the province's quality of accommodation as compared with other provinces. The lodging facilities in Nueva Ecija have been structured based on the nature-based peripheral attribute. Most destinations constructed accommodation into an upscale experiential lodging such as glamping (glamorous camping) and villas, among others. Nueva Ecija has satisfied destination quality in terms of cleanliness. A clean tourist environment associates in making a good first impression, an apprehension that the destination preserves its natural features, and a good depiction conveying environmental conservation. Moreover, the quality of public infrastructure, reception services, communication, level of safety, and availability of shopping facilities were identified as significant factors contributing to the tourism value chain. Stakeholders responded with a need to improve the quality of public facilities such as toilets, parking areas, recreation parks, and picnic grounds.

Based on the document review, management plan and tourism regulations were made available; however, not all destinations properly implemented and communicated these with the local community residents participating in the operations. Proper communication among internal stakeholders could rationalize the conveyance of information and implementation. Despite not having specialized tour guides in Nueva Ecija ecotourism destinations, residents function as local guides to tour and orient tourists with the destination. Local government unit (LGU) and other relevant government agencies support in uplifting the knowledge and skills of residents through training, workshops, and livelihood programs. However, some of these government endeavors were not sustained as key informants viewed with the lack of monitoring and evaluation.

A characteristic and emotionally interesting destination personality has been constructed to leverage the distinguished image of a destination and influence tourists' selection behavior. With the objective of being exceptional, Nueva Ecija portrays a provincial community aspect which sets aside the state of being glamorous but instead captivates the market of unique personalities such as the province's charismatic appeal to relaxation while enjoying and experiencing the authentic Novo Ecijano life.

With the promotion of Nueva Ecija tourism, the provincial tourism office of the province launched a mobile application to showcase the tourist destinations, products, services, culture, and heritage of Nueva Ecija with the slogan of "Never Ending" highlighting letters N and E which represent Nueva Ecija. This has been a good initial move to communicate with larger markets and create bigger demand in tourism. However, as associated with destination quality, not all internal and external stakeholders were aware of this endeavor. Stakeholders also shared their sentiments that this endeavor must be communicated to all tourism sectors and stakeholders. Thus, building community awareness stretches out to communicate a sense of ownership and responsibility to own locality. Awareness can be facilitated in a form of education starting at the community level (Chan, 2021). Nature and culture motivations set off valuable constructs in defining destination branding of a destination. The purpose of destination culture is to represent the core values of the destination's distinct culture, local geography, and traditions including events and festivals. Relevant with this, with the abundance of the destination on its natural resources, the province may also leverage cultural aesthetic potential through capacitating the local community in developing indigenous crafts.

SWOC Analysis Vis-à-vis Stakeholder Driven Branding of Nueva Ecija as an Ecotourism Destination

In Susanto et al. (2022), SWOT analysis has been utilized as a strategic planning method in evaluating the strengths, weaknesses, opportunities, and threats of a tourism destination and in developing external and internal factors toward tourism development. Similarly, the constructs of destination branding gathered from pertinent documents among Nueva Ecija tourism-related offices (i.e., the provincial tourism office, tourism offices of LGUs and Department of Environment and Natural Resources [DENR]) have been carefully analyzed and evaluated with in-depth perspectives to outline definite internal factor analysis (IFAS) (strengths and weaknesses) and external factor analysis (EFAS) (opportunities and challenges) to further clarify current conditions and potential strategies on the initiation and implementation of destination branding (Susanto et al., 2022).

Strengths

The tourism industry of Nueva Ecija has been assessed in Vigilia (2020) which statistically outlined the competitive advantage of the province in terms of tourist spots' appeal, accessibility, and availability of products and services. Considering the characteristic appeal of tourists' spots in the province, Nueva Ecija showcases its uniqueness to other province's tourists, where the province was able to preserve its nature beauty or natural aesthetic with a portion of modern-day enhancement; tourism activities have been added as part of value addition of the physical aesthetics where tourists may also engage with various nature activities while enjoying the scenery.

The province also showcases historical and cultural values in terms of preserving historical tourist spots over years of existence and in preserving traditions and customs of local inhabitants, respectively. The province of Nueva Ecija has issued Provincial Ordinance No. 37 series of 2020 for the preservation, conservation, and protection of the province's cultural properties, historical landmarks, shrines, and monuments. This is in accordance with the National Heritage Act of 2009 which mandates the LGUs to safeguard and preserve heritage sites, register cultural properties, and preserve traditional arts noting these as important tangibles and intangible aspects of heritage and as prime mover of cultural diversity and sustainable development.

In the Tourism Development Plan established for the LGUs in Region III, tourism has been defined as a source of private profit and economic growth. Parallel with the previous provisions of showcasing culture and heritage in the province, tourist destinations provide mutual benefit to the host community – preserving natural aesthetics and providing employment and livelihood privilege to the local community.

According to the Regional Director of DOT Region III, visiting destinations which do not degrade the quality of the environment captured a greater demand because tourists have elevated their sense of social and environmental responsibility. A margin of 40%–60% of tourists chooses nature-based destinations which offer a wide range of authentic environmental and cultural experiences. Relevant with this, among these, "ecotourists" in the Philippines showed that they were willing to spend more than $50 (approximately 2,944.00 pesos) per trip to help conserve the natural areas. In Aguilar and Escalera-Reyes (2020), the provincial market has been distinguished as not too demanding when it comes to the standard of services and amenities and prefer to spend for an accommodation ranging from an ideal minimal rate between 1,000 and 2,000 pesos per night and for entrance fees between 100 and 200 pesos. Providing affordable tourism experiences captures more potential markets.

Ensuring security, public order and safety were set as high regard in the Central Luzon Regional Development Plan. Region III is envisioned to have a better capability to ensure the safety and security of its citizenry from all forms of criminality and illegal drugs. Central Luzon has organized that all barangays in the region have Barangay Peacekeeping Action Teams (BPATs); this has been evaluated as an efficient movement which controlled the increase of crime rate in the region. BPATs have been a dynamic and proactive approach to community peacekeeping that Philippine National Police (PNP) pushes for. Its security and safety services conventionally distributed by the PNP were improved by the composition of the following: (a) barangay *tanod*, (b) security guards, (c) anti-crime nongovernmental organization (NGO) members, (d) sectoral group members, (e) tricycle operators/drivers association members, and (f) radio groups/club members. Aside from this, the province of Nueva Ecija has intensified efforts in reducing criminality as well as established reliance management plan among protected areas with the help of DENR – making local community properly guided with an ecotourism management plan and that the ecotourism destinations will be prepared and resilient for unpredictable risks and hazards. This strategy promotes a safe tourist environment for everyone.

Weaknesses

There were still various stakeholder issues and concerns with the transportation and infrastructure such as transfer connectivity, access, destination, and site infrastructures including parking, public toilets, information and restoration.

In the LGU Tourism Guidebook, part of the availability of onsite facilities is to enhance the destination's accessibility with directional and information signage. Since most of the nature destinations in the province were situated in mountains and in village areas, there should be adequate signages. Furthermore, the ecotourism destinations in the province lack proper visitor management that includes a designated briefing area to inform tourists prior to enjoyment of the facilities has been considered as an initial point in the visitor flow. Proper orientation would help tourists be familiar with the destination, including familiarity with the rules and regulations.

In Alejandria-Gonzalez (2016), marketing plans may be strategized while highlighting authenticity and diversity of cultural products. Most destinations have utilized social media to advertise and reach more potential markets in their destinations. However, these destinations also look for the initiatives from the LGUs or the provincial government to support them in strategizing with promotions and marketing. Advertising the scenery of the destinations and including promotional packages with culture and heritage is a platform recognized on the tourism development plan issued by the regional provisions.

Opportunities

The National Tourism Development Plan (NTDP) outlined strategic directions focusing on improving competitiveness and enhancing growth and pursuing sustainability and inclusive growth. Listed under the pursuance of sustainable and inclusive growth are support to micro, small, and medium enterprise development, gender, and women empowerment initiatives, expand and promote cultural offerings, adopt measures to enhance environment protection and climate change adaptation, and formulate measures to address tourism risks and crisis.

The province of Nueva Ecija is rich with both agritourism and ecotourism resources; these natural resources were also the livelihood basket of the local communities. The province's local produce may be enhanced or developed with the help of research and technology and with partnership to various institutions. LGUs provided free training to the local communities such as tour guiding, cookery, and housekeeping services; these programs aim to capacitate the human resource, making the local residents more qualified in participating in the destination's operations. Other workshops conducted for the women community are craft making and culinary using local produce. The LGU also provided start-up capital to the women community to start their business. In the coming years and part of the regional development plan, more job opportunities will also be made available adjacent with the development of these destinations.

Challenges

The NTDP envisioned developing a highly competitive and environmentally sustainable tourism industry that is focused on creating inclusive growth adopted to provide the long-term implementation framework consistent with the declaration of policy set forth in the Tourism Act of 2009. National and regional provisions set various parameters in enhancing the products and services, capacitance of the human resource, and preserving and safeguarding the natural resources. Opportunities were perceived concurrent with the state of the destination, yet there were still recognized limitations of these destinations and regard to their operations.

In Bautista (2021), it has been emphasized that a socially responsible community can be achieved through various programs for the community to have initial knowledge on protecting, managing, and conserving the natural assets through a sense of ownership. The LGUs provided livelihood training and workshops to the local community and granted starting capital investment for a small business which could be craft making or culinary. However, no monitoring schemes facilitated whether the businesses have been sustained or gathered feedback and evaluation from the community.

The LGU tourism development plan highlighted the need to leverage the natural and socioeconomic environment. However, regardless of the development plans and government mandates, there were still unpredictable instances or weather conditions which cannot be controlled such as the pandemic, climate change, and natural disasters. Inflation is another recognized threat to these destinations. Despite offering products and services at provincial rates, because of inflation, the cost of goods and services will also increase.

Too much development in the area may also pose a threat, especially for the residents including the indigenous peoples. Most of the destinations provided livelihood and employment opportunities to the local community; however, they tend to be tourism dependent because of seasonality.

From the result derived from analysis of Internal Factor Analysis – External Factor Analysis (IFAS-EFAS) matrixes, the next procedure has analyzed the statistical figures with the use of a SWOT analysis diagram to determine the most applicable strategy using the Adaptive Analysis Method in Prodjonoto (2020) (i.e., growth, diversification, turn around, and defensive). Using the IFAS and the EFAS calculation in Table 12.1, the computed set of weighted value scores for strengths component was 3.15, weaknesses component was 2.85, opportunities component was 3.25, and threats component was 2.95.

The value of IFAS has been calculated by subtracting the total weighted value of weaknesses component from the strength's component, while using the similar calculation for the external components.

Table 12.1. IFAS and EFAS SWOC Factors and Coordinate Calculations.

No	*Strengths*	*Weight*	*Value*	$W \times V$
1	Beautiful scenery	0.25	3	0.75
2	Abundant homegrown produce	0.25	3	0.75
3	Affordable price	0.15	3	0.45
4	Cultural and heritage attractions	0.20	3	0.60
5	Safe and clean tourist environment	0.15	4	0.60
	Total Strengths Factor	1.00		3.15
No	*Weaknesses*	*Weight*	*Value*	$W \times V$
1	Stakeholders' lack of awareness or involvement in tourism promotion	0.25	3	0.75
2	Access and signage to the tourist destination are inadequate	0.20	3	0.60
3	Lack of proper visitor management	0.15	2	0.30
4	Lack of transportation services	0.20	3	0.60
5	Lack of marketing strategies	0.20	3	0.60
	Total Weaknesses Factor	1.00		2.85
No	*Opportunities*	*Weight*	*Value*	$W \times V$
1	Increasing market demand for nature travel	0.15	3	0.45
2	Product development	0.20	3	0.60
3	External support from government agencies, NGOs, academe, and private entities	0.25	3	0.75
4	Livelihood and employment opportunities to local community	0.25	4	1.00
5	Enhancement of tourist destination and facilities	0.15	3	0.45
	Total Opportunities Factor	1.00		3.25
No	*Threats*	*Weight*	*Value*	$W \times V$
1	Continuity of external support	0.25	4	1.00
2	Gentrification/displacement of local inhabitants	0.25	3	0.75
3	Market seasonality	0.15	2	0.30

(Continued)

Table 12.1. *(Continued)*

4	Inflation/increase in the level of prices of goods and services	0.20	3	0.60
5	Too much development in the area	0.15	2	0.30
	Total Threats Factor	1.00		2.95
	Description	*Calculation*	*Coordinate Value*	
IFAS	Strengths factor – Weaknesses factor	3.15 − 2.85	$x = +0.30$	
EFAS	Opportunities factor – Threats factor	3.25 − 2.95	$y = +0.30$	

Source: Author constructed.

Further, by mapping the IFAS value on the horizontal (x-axis) and the EFAS value on the vertical (y-axis) of a two-dimensional Cartesian coordinates diagram (Setiadi et al., 2021), it can be determined that the best strategy is in Quadrant I (Growth Strategy) as both x and y values are determined to be positive, as shown in Fig. 12.1. After computing the total of strengths, weaknesses, opportunities, and threats in terms of IFAS (value of strengths deducted by the value of weaknesses),

Fig. 12.1. IFAS/EFAS Strategy Coordinate. *Source:* Author constructed based from Prodjonoto (2020) and Ritonga et al. (2018).

which is 3.15 − 2.85 = + 0.30, and in terms of EFAS (value of opportunities deducted by the value of threats), which is 3.25−2.95 = + 0.30; it was mapped that the strategy coordinate on horizontal axis and vertical axis resulted both in a positive value. As illustrated in Fig. 12.1, result corresponds to Quadrant I strategy, in which the province of Nueva Ecija may consider in taking advantage of utilizing the Growth Strategy to develop the province's ecotourism destination branding.

Conclusion

Destination image of Nueva Ecija has been perceived as an all-encompassing concept among all stakeholders. Stakeholders suggested a commonality of perceptions with the nature-based tourism value assets of the province. The natural and local resources were all made available; however, stakeholders must work toward the same goals of enhancing the products and services, augmenting the ecotourism destinations prepared in the postpandemic, capacitating the knowledge and skills of human resource, and empowering local residents in contributing in the operations and progressing the local resources including the enhancement of livelihood and well-being of local community.

Creating a destination brand identity for the province using the destination's unique characteristics and activities as means to be differentiated and distinguished itself from other destinations may serve sustainable advantage in positioning the destination or differentiating a destination despite few similar attributes to nearby ecotourism destinations. Positioning can be strategized by the LGU by recognizing the tangible and intangible attributes of their ecotourism products. Investing on logo, quality of design, marketing tools such as depictions of the location, the natural resources, souvenirs, and handicrafts may establish the tangible attributes of the ecotourism destinations in the province, while intangible attributes such as effect to the emotions of tourists, tourist interaction, and appeal to the senses are key to advance the destination's intangible components.

Two (2) basic components have been identified to measure tourist destination quality: technical (measurable elements) and functional (extent how services were provided). Aligned with these components, destination quality dimensions were composed of various indicators in attraction, people (providing service), facilities available, quality of infrastructure, and tourist safety. Nueva Ecija's ecotourism destinations must drive the quality of infrastructure facilities particularly accessibility, transport services, pricing, safety, and convenience. These indicators have been highlighted with substantial contribution to destination quality.

The ecotourism destinations in the province showcase a circuit of promising tourist destinations. Yet, potential must be maximized toward longevity and sustainability; developing tourist destinations is a continuous process influenced by internal and external factors contributing to better market position. Collaboration among internal and external stakeholders has also been highlighted of significance; thus, networking and linkages among the tourism service providers, government agencies, and private entities have to be strengthened to penetrate opportunities and linkages to leverage tourist services and product development.

Despite the drawbacks, the global scenario had served as a wakeup call for the province regarding the possibility of experiencing risk which has been determined that could destroy or damage assets and even cause life loss. With this, the province has to be strategically prepared and resilient. Destination resilience is a call for collaboration among internal and external stakeholders toward the movement of a sustainable and resilient tourist environment. Thus, a safe and secure tourist environment is adjacent with the reliability of the destination quality.

Another perceived response in management strategy is uplifting the livelihood condition of the local community through magnifying indigenous product offerings and career training in furtherance of their knowledge and skillset. Additionally, the ecotourism destinations in the province may strategically accomplish its branding anchored with the destination branding strategy framework.

In Bautista (2021), LGU and other relevant government agencies initiatives on conducting training and grant livelihood programs must be monitored to be successful. Thus, the study highlights that in evaluating and monitoring a tourist destination, the focus is not only on the aperture of tourist demand and feedback but also on the status of the local community. The residents in the ecotourism sites primarily depend on their livelihood and employment with the tourist destination. These locals should be allocated with contingency planning to cope with seasonality and avoid too much dependence on the tourist destination.

The province may leverage from marketing strategy aligned with the stakeholder-driven branding of Nueva Ecija as an ecotourism destination. In branding a destination, correct representation must be addressed in between missing links of what has been communicated, perceived ideal, and desired perspectives which can be done through market research and analysis or feasibility study.

Feedback could be a critical juncture in business, so as in managing tourism destinations. Reviews and word-of-mouth (WOM) could make or break a business, as not all WOM yields positive responses. Yet, feedback and evaluation help a business to further strategize some lapses from tourist entry point to departure. Feedback serves as the evaluation of the "moment-of-truth" – whether tourists have been satisfied or dissatisfied. Monitoring, on the other hand, provides an opportunity to review the phases of the destination branding strategic framework.

Tourism destinations across the globe realized that unpreparedness to withstand a catastrophe is inevitable. However, despite the downsides during the onset of the pandemic, the province of Nueva Ecija ascended and discovered a new treasure in ecotourism. The province unraveled agritourism as a catalyst to engage resources in ecotourism and coping mechanisms during the pandemic. Tourism industry has been hardly hit by COVID-19; yet tourism also lifted the economic condition of the province during the pandemic until today. Tourism destinations in the country and neighboring countries in the ASEAN may learn from these takeaways of the province. Critical planning, formulation of contingencies, and market-driven reorganizations toward branding are noteworthy approaches to recoil from unforeseen challenges were evidently some of the crucial approaches employed by ecotourism stakeholders in the province. One of the destination's distinct features is responding to the market needs and demands. To successfully position a destination, it must be aligned with the tourists' need attributed to their sociodemographic profile, preferences, trends, and future demands. Identifying their market segmentation may also serve as a tactical instrument in gathering a definite representation among heterogeneous potential tourist groups. This strategic tool in identifying the market helps tourism service providers in aligning products and services and in allocating marketing strategies properly.

References

Aguilar, A. L. D., & Escalera-Reyes, J. (2020). Family relations and socio-ecological resilience within locally-based tourism: The case of El Castillo (Nicaragua). *Sustainability*, *12*(15), 5886. https://doi.org/10.3390/su12155886

Alejandria-Gonzalez, M. C. P. (2016). Cultural tourism development in the Philippines: An analysis of challenges and orientations. *Journal of Quality Assurance in Hospitality & Tourism*, *17*(4), 496–515. https://doi.org/10.1080/1528008X.2015.1127194

Aruta, J. J. B. R., & Guinto, R. R. (2022). Climate anxiety in the Philippines: Current situation, potential pathways, and ways forward. *The Journal of Climate Change and Health*, *6*(1), 100138. http://dx.doi.org/10.1016/j.joclim.2022.100138

Bansil, P. D. D., Capellan, S. A. R., Castillo, R. C., Quezon, C. D., & Sarmiento, D. M. B. (2015). Local community assessment on the economic, environmental and social aspects of ecotourism in Lobo, Philippines. *Asia Pacific Journal of Multidisciplinary Research*, *3*(4), 132–139. https://www.apjmr.com/wp-content/uploads/2015/10/APJMR-2015-3.4.3.17.pdf

Barr, J. C., Woosnam, K. M., Henn, K. A., & Andrews, C. T. (2021, June 30). Support for nature-based tourism during the pandemic: Considering Bohol. In *Asia Pacific tourism association conference 2021*. Busan.

Bautista, M. J. D. (2021). An ecotourism development plan for Minalungao National Park, General Tinio, Nueva Ecija. *International Journal of Advanced Multidisciplinary Studies*, *1*(3). https://www.ijams-bbp.net/wp-content/uploads/2021/07/Marie-Jel-D.-Bautista.pdf

Beall, J. M., Boley, B. B., Landon, A. C., & Woosnam, K. M. (2021). What drives ecotourism: Environmental values or symbolic conspicuous consumption? *Journal of Sustainable Tourism*, *29*(8), 1215–1234. https://www.tandfonline.com/doi/full/10.1080/09669582.2020.1825458

Chan, C. S. (2021). Developing a Conceptual Model for the Post-COVID-19 Pandemic Changing Tourism Risk Perception. *International Journal of Environmental Research and Public Health*, *18*(18), 9824. https://doi.org/10.3390/ijerph18189824

Chin, C. G. R., Cuasay, D. A. N., Galvez, M. K. D., Tamares, J. C. B., & Oreta, M. S. L. (2021). A study on the travel preferences of millennials in the City of Manila for post-pandemic domestic travel. *AVAHAN: A Journal on Hospitality & Tourism*, *9*(1), 46–63.

Choi, Y. G., Ok, C., & Hyun, S. S. (2011). *Evaluating relationships among brand experience, brand personality, brand prestige, brand relationship quality, and brand loyalty: An empirical study of coffeehouse brands*. https://api.semanticscholar.org/CorpusID:19188766

Clark, G., Feldman, M., & Gertler, M. (2000). *Economic geography: Transition and growth*. Oxford University Press. https://ora.ox.ac.uk/objects/uuid:9e099aae-fb91-4bfd-b323-aa8d1f7d7e6e

Cobbinah, P. B. (2015). Contextualizing the meaning of ecotourism. *Tourism Management Perspectives*, *16*, 179–189. https://doi.org/10.1016/j.tmp.2015.07.015

Department of Tourism. (2011). *A strong Philippine tourism in 2011: Department of Tourism Year-End Report*. http://tourism.gov.ph/files/2011%20DOT%20Year%20End%20Report.pdf

Desai, B. H. (2022, September). 14. United Nations Environment Program (UNEP). *Yearbook of International Environmental Law*. https://doi.org/10.1093/yiel/yvac039

Fernando, I. N. (2015). What competitive strategies way forward the regional competitiveness? A comparative economic approach to Sri Lankan tourism. *International Journal of Business and Management*, *10*(4), 178. https://doi.org/10.5539/ijbm.v10n4p178

Fung, T., & Wong, F. K. (2007). Ecotourism planning using multiple criteria evaluation with GIS. *Geocarto International*, *22*(2), 87–105. https://doi.org/10.1080/10106040701207332

Gutierrez, E. L. M. (2021). Redefining sustainable tourism in COVID-19: A political-economy approach. *DLSU Business and Economics Review*, *30*(1). https://www.dlsu.edu.ph/wp-content/uploads/2021/04/1Gutierrez-033021.pdf

Ivanova, M., Ivanov, I. K., & Ivanov, S. (2021). Travel behaviour after the pandemic: The case of Bulgaria, Anatolia. *An International Journal of Tourism & Hospitality Research*, *32*(1), 1–11. https://doi.org/10.1080/13032917.2020.1818267

Konecnik, M., & Go, F. (2008). Tourism destination brand identity: The case of Slovenia. *Journal of Brand Management*, *15*(3), 177–189. https://link.springer.com/article/10.1057/palgrave.bm.2550114

Konecnik, M., & De Chernatony, L. (2012). Developing and applying a place brand identity model: The case of Slovenia. *Journal of Business Research*, *66*(1). https://www.researchgate.net/publication/256043355_Developing_and_Applying_a_Place_Brand_Identity_Model_The_Case_of_Slovenia

Kusumaningrum, D. A., & Wachyuni, S. S. (2020). The shifting trends in traveling after the COVID 19 pandemic. *International Journal of Tourism & Hospitality Reviews*, *7*(2), 31–40. https://doi.org/10.18510/ijthr.2020.724

Mayo, F. L., Maglasang, R. S., Moridpour, S., & Taboada, E. B. (2021). Exploring the changes in travel behavior in a developing country amidst the COVID-19 pandemic: Insights from Metro Cebu, Philippines. *Transportation Research Interdisciplinary Perspectives*, *12*, 100461. https://doi.org/10.1016/j.trip.2021.100461

Ocampo, L., Ebisa, J. A., Ombe, J., & Escoto, M. G. (2018). Sustainable ecotourism indicators with fuzzy Delphi method – A Philippine perspective. *Ecological Indicators*, *93*, 874–888. https://doi.org/10.1016/j.ecolind.2018.05.060

Pader, L. D., Banson, C. J. G. C., & Angelou, W. (2017). DNA barcoding of Chiropterans at Minalungao National Park, Nueva Ecija, Philippines. *International Journal of Agricultural Technology*, *13*(7.3), 2341–2344. https://www.cabidigitallibrary.org/doi/pdf/10.5555/20183304221

Padin, M. G. (2016). *Tourism sector sustains upward trajectory*. Philippine Star. https://www.philstar.com/business/2016/12/30/1657893/yearender-tourism-sector-sustains-upward-trajectory

Pagoso, C. M., & Katigbak, P. (2009). Ecotourism development plan at Pantabangan, Nueva Ecija: A survey of community residents. *LCCM Research Journal*, *19*(1). https://ejournals.ph/article.php?id=10077

Pestek, A., & Nikolic, A. (2011). Role of traditional food in tourist destination image building: Example of the city of Mostar. *UTMS Journal of Economics*, *2*(1), 89–100. https://utmsjoe.mk/files/Vol.%202%20No.%201/0-2-1-10-Pestek-Nikolic.pdf

Pike, A., Dawley, S., & Tomaney, J. (2010). *Resilience, adaptation and adaptability*. ResearchGate; Oxford University Press. https://www.researchgate.net/publication/46512980_Resilience_Adaptation_and_Adaptability

Prodjonoto, K. W. A. (2020). Introducing adaptive analysis methods: The present analysis & future analysis. *Journal of Analytical Sciences, Methods and Instrumentation*, *10*(2). https://www.scirp.org/journal/paperinformation?paperid=101625

Ritonga, H., Nasrudin, S., Miftah, E. F., Ritonga, C. P. M., Tharmizi, H., Maya, S., Yossie, R., & Nasution, M. D. T. P. (2018). Rural tourism marketing strategy and swot analysis: A case study of Bandar Pasir Mandoge sub-district in north Sumatera. *International Journal of Civil Engineering and Technology*, *9*, 1617–1631. https://www.researchgate.net/publication/328552386_Rural_tourism_marketing_strategy_and_swot_analysis_A_case_study_of_bandar_pasir_mandoge_sub-district_in_north_sumatera

Rivera, J. P. R., & Gutierrez, E. L. M. (2018). A framework toward sustainable ecotourism value chain in the Philippines. *Journal of Quality Assurance in Hospitality & Tourism*, 1–20. https://doi.org/10.1080/1528008x.2018.1492495

Robinson, J. (2020). Thoughts on the post-pandemic new normal in air travel. *Journal of Airport Management*, *15*(1), 6–19. https://ideas.repec.org/a/aza/jam000/y2020v15i1p6-19.html

Roxas, F. M. Y., Rivera, J. P. R., & Gutierrez, E. L. M. (2020). Mapping stakeholders' roles in governing sustainable tourism destinations. *Journal of Hospitality and Tourism Management*, *45*, 387–398. https://doi.org/10.1016/j.jhtm.2020.09.005

Seakhoa-King, A., Augustyn, M. M., & Mason, P. (2020). *Tourism destination quality: Attributes and dimensions*. https://www.emerald.com/insight/publication/doi/10.1108/9781839095580

Setiadi, A., Rudwiarti, L., Priscilia, F., & Wardhani, M. (2021). City tourism branding resilience during the COVID-19 pandemic in Yogyakarta, Indonesia. *Spatium*, 1–8. https://doi.org/10.2298/SPAT2145001S. https://www.researchgate.net/publication/354069312_City_tourism_branding_resilience_during_the_COVID19_pandemic_in_Yogyakarta_Indonesia/citations

Shariffuddin, N. S. M., Zain, W. M. A. W. M., & Azinuddin, M. (2020). Collaborative challenges among stakeholders on tourism destination competitiveness. *International Journal of Innovation, Creativity and Change*, *13*(1), 454–466. https://www.ijicc.net/images/vol_13/13138_Shariffuddin_2020_E_R.pdf

Susanto, E., Hendrayati, H., Rahtomo, R. W., & Prawira, M. F. A. (2022). Adoption of digital payments for travelers at tourism destinations. *African Journal of Hospitality, Tourism and Leisure*, *11*(2), 741–753. https://www.ajhtl.com/uploads/7/1/6/3/7163688/article_14_11_2_741-753.pdf

Tsaur, S. H., Yen, C. H., & Yan, Y. T. (2016). Destination brand identity: Scale development and validation. *Asia Pacific Journal of Tourism Research*, *21*(12), 1–14. http://dx.doi.org/10.1080/10941665.2016.1156003

Viernes, E. J. Y., & Aquino, M. U. (2020). Agri-tourism as strategic intervention for tourism education: Lessons learned from Central Luzon State University, Philippines. *Journal of Hospitality and Tourism Issues*, *2*(2), 161–177. https://dergipark.org.tr/en/download/article-file/1228090

Vigilia, A. Y. (2020). Assessment of tourism industry in Nueva Ecija. *International Journal of Advanced Engineering, Management and Science*, *6*(1). https://dx.doi.org/10.22161/ijaems.61.6

Yoo, J. W., Park, J., Lee, J. H., & Park, H. (2022). Recovering from the COVID-19 shock: The role of risk perception and perceived effectiveness of protective measures on travel intention during the pandemic. *Service Business*, 1–24. https://doi.org/10.1007/s11628-021-00476-3

Ziegler, J., Araujo, G., & Labaja, J. (2021). Can ecotourism change community attitudes towards conservation? *Oryx*, *55*(4), 546–555. https://www.cambridge.org/core/journals/oryx/article/can-ecotourism-change-community-attitudes-towards-conservation/7B83C9848D8872256B74116E380FC2AF

Chapter 13

Revisiting the Regulatory Function of Government Toward Tourism Sustainability and Resilience: Basis for Policy Formulation

John Paolo R. Rivera[a,b] *and Warner M. Andrada*[c]

[a]Philippine Institute for Development Studies, Philippines
[b]Asian Institute of Management – Dr. Andrew L. Tan Center for Tourism, Philippines
[c]Department of Tourism, Philippines

Abstract

While government is known to provide political guidance and exercising its executive function, it is also has regulatory powers through laws it enacts. In fostering sustainability, it is important to inquire how government's role can be innovated to facilitate sustainability, particularly in the travel and tourism industry. By reviewing tourism governance literature and mapping governance roles in the travel and tourism industry, this chapter creates a policy framework that underscores a new approach to tourism governance. We underscore that government's role must pivot toward being more developmental than regulatory so that it can effectively stimulate the market to sustainability by fostering value creation, supporting manpower capacitation, ensuring health and safety, and protecting the environment. This can be done if government will not fix the market and promote free market policymaking.

Keywords: Development; government; market; policy; regulation; sustainable tourism

Revisiting Sustainable Tourism in the Philippines, 237–250

Published under exclusive licence by Emerald Publishing Limited
doi:10.1108/978-1-83753-678-820241013

Introduction

Government has a mandate that provides political guidance, exercises its executive function, and implements regulations supported by the laws that it enacts (Danziger, 2012). These are provided for and governed by a Constitution (Fabbri & Faure, 2018). In the tourism sector, because governance is the process by which groups of people make decisions (Graham et al., 2003), government's role is profound as it has the capacity to develop policies that will enhance a particular destination. It offers major potential for the management of tourism destinations by exploring the constituents of tourism destinations and focusing on providing direction and boundaries for destinations (Bichler, 2021). Hence, as argued by Bhuiyan et al. (2023), good governance can promote sustainable tourism development.

In the Philippines, the tourism industry is primarily governed by the Department of Tourism (DOT).[1] It is the Philippine government's executive department whose mission is to "encourage, promote, and develop tourism as a major socio-economic activity to generate foreign currency and employment and to spread the benefits of tourism to both the private and public sector" (DOT, n.d., para. 1). Its vision is "to develop a globally competitive, environmentally sustainable and socially responsible tourism industry that promotes inclusive growth through employment generation and equitable distribution of income thereby contributing to building a foundation for a high trusted society" (DOT, n.d., para. 2). From these, two things are apparent: it is a government agency that is mandated to (1) regulate the country's tourism industry and (2) promote the country as a tourist destination. This regulatory function of the government, through DOT, has also been underscored by the frameworks designed by Roxas et al. (2020b), Haiying et al. (2018), Koscak and O'Rourke (2017), Erkuş-Öztürk (2011a, 2011b), Clifton and Amran (2010), Byrd (2007), Provan and Kenis (2007), Miller and Twining-Ward (2005), Björk (2000), Bramwell and Lane (2000), Fennell and Malloy (1999), and Buhalis and Fletcher (1995).

Government's regulatory role in tourism has been emphasized by Roxas et al. (2020b) in their "deconstructed star model of tourism stakeholders" (p. 393) wherein the "synergy between tourism authorities comprised of international bodies, national government, and local government enables them to do regulation" (p. 392). Moreover, Luo and Zhang (2015) argued that tourism regulations are developed for socioeconomic, environmental, and political purposes and are enforced by various institutions, including national and local governments, private professional bodies, and nongovernment organizations – all of which creates a governance structure.

However, with the hard lessons tourism learned from the COVID-19 pandemic (Gössling & Schweiggart, 2022), it prompted a change in governance structure and a pivot in the involvement of government in tourism (Allaberganov et al., 2021; Babii & Nadeem, 2021; Robina-Ramirez et al., 2022; Wan et al., 2022). This chapter aims to explicate a new and revitalized role of government in the governance structure of tourism. To address this objective, we need to answer the

---

[1]Official website: https://beta.tourism.gov.ph/

point of inquiry *how the government can form the market toward sustainable tourism without justifying policy?*

As such, we are able to craft an innovative approach to tourism governance resulting to a recalibrated policy framework toward sustainable tourism particularly for a less developed country like the Philippines. In pursuing sustainable tourism, policy frameworks cannot be confined to just regulation and fixing the market (e.g., Boracay Rehabilitation Program of 2018 when government fixed the market to enhance Boracay as a product, improve tourists' total experience through activities, facilities, and transportation) but must also be anchored on a multifaceted agenda emerging from both international (e.g., United Nations Sustainable Development Goals [UN SDG], ASEAN Framework on Sustainable Tourism Development, UN Tourism Framework on Measuring Sustainability in Tourism) and national (e.g., Philippine Development Plan [PDP 2023-2028], National Tourism Development Plan [NTDP 2023-2028] of the Philippines) frameworks.[2–7]

All share intertwined goals of facilitating the financial viability and economic sustainability of tourism resources; increasing the overall capacity of tourism stakeholders to implement sustainable development plans within their jurisdictions; improving coordination and collaboration among stakeholders across the tourism value and supply chains (Ammirato et al., 2014; Graci, 2013); creating more profit-oriented opportunities that benefit enterprises, communities, and the environment; and ensuring that economic, social, and environmental impacts of tourism are beneficial for everyone.

This chapter emanates from the study of Rivera et al. (2022) calling for the mobilization of the principles of sustainability, stakeholder participation, and safe travels to rethink, reimagine, and redesign existing frameworks to enhance tourism governance to reinforce visitor's confidence toward travel, thereby creating a sustainable, safe, and resilient postpandemic tourism industry (Asia-Pacific Economic Cooperation [APEC], 2024).

Government's Revitalized Postpandemic Role in Tourism

In revitalizing government's role in a postpandemic travel and tourism industry, we recall the argument of Hall (2005) that government's role in regulating tourism

---

[2]For more information, see https://www.denr.gov.ph/images/DENR_Accomplishment_Report/Boracay_Rehabilitation_Program.pdf

[3]For more information, see https://sdgs.un.org/goals

[4]For more information, see https://asean.org/wp-content/uploads/2023/01/ASEAN-Framework-on-Sustainable-Tourism-Development_compressed.pdf

[5]For more information, see https://www.unwto.org/tourism-statistics/measuring-sustainability-tourism

[6]For more information, see https://pdp.neda.gov.ph/wp-content/uploads/2023/01/PDP-2023-2028.pdf

[7]For more information, see https://beta.tourism.gov.ph/news_and_updates/dot-chief-unveils-national-tourism-development-plan-ntdp-2023-2028-at-stakeholders-summit/

is not only anchored on the various mandates of the state but also on specific management issues such as coordination and kind of intervention.

According to Roxas et al. (2020b), national governments design policy frameworks driven by the emergence of movements toward public–private partnerships (PPP) and networks. This stimulates the "self-governance" capacity of local governments and businesses by partaking in economic activities, including tourism (p. 393). Because of this response to PPP and networks, local government becomes the prime movers of the interests of local communities (Tosun, 2006). Hence, the national government is deemed to have a vital role in regulating the system to which tourism development occurs (Bevir, 2009).

Meanwhile, multilateral organizations such as the UN Tourism (formerly United Nations World Tourism Organization [UNWTO]) have spearheaded a form of global governance through its code of ethics on tourism.[8] Likewise, through the Global Sustainable Tourism Council (GSTC), a multi-stakeholder international tourism organization has also influenced the way sustainability is being pursued (Bushell & Bricker, 2016).[9] While the UN Tourism and GSTC have no regulatory force, they can influence national governments to align (Jamal & Camargo, 2014).

However, due to the evolving landscape of tourism governance (Hall, 2005), the UN Tourism has been formulating tourism laws anchored on sustainable development, management, and regulation of the industry by identifying the rights and responsibilities of tourism stakeholders (i.e., national governments, enterprises, consumers), in the sustainable utilization of natural and cultural resources toward development (UNWTO, 2013).

Redefining Government's Regulatory Role

In enabling government's regulatory function to support the renewed demands of the travel and tourism industry in a postpandemic world, this regulatory function must be redefined. Because the travel and tourism industry is an economic pillar for many countries (Gutierrez, 2021), it warrants government control as it provides goods and services through its profitable enterprises to protect the rights of the demand- and supply-side stakeholders (Agafonova & Spektor, 2023), as well as the preservation and conservation of the natural environment (Roxas et al., 2020a).

Because both demand- and supply-side of the travel and tourism industry has evolved (Brouder et al., 2017; Gowreesunkar et al., 2024; Rivera et al., 2022; Rodolfo et al., 2022) given the trajectory of sustainability discourse, impacts of climate change, call for good governance, among others, government must go beyond the usual regulatory function (e.g., accreditation, certification, recognition, market surveillance, fixing the market, among others) (Andrews et al., 2021; Organization for Economic Co-operation and Development [OECD], 2020). While these are still important and meant to remain, it is now imperative to have

[8]Official website: https://www.unwto.org/

[9]Official website: https://www.gstcouncil.org/

a mindset that views regulation not as a form control but as a facilitating factor of sustainability (Atalla et al., 2022; Rakhmawati et al., 2020) and resilience (OECD, 2020; O'Grady & Shaw, 2023). For example, because the market is now self-regulating (Whitworth, 2023), government can intervene through human capacity development to contribute to sustainability and resilience.

On the supply side, capacitating tourism manpower is important because competition has also evolved driven by technology that changed tourism demand (Lee et al., 2021; Ray, 2020). The pandemic has also changed the landscape of travel and tourism (DOT et al., 2020, 2021). There are existing business models that have become obsolete, thereby needing adjustments and reassessments (Gutierrez et al., 2020). Hence, a new and different market emerged and competed with the traditional market we know prior to the pandemic. As the pandemic caused havoc on the travel and tourism industry (Rivera et al., 2022; Roxas et al., 2022), which is primarily composed of micro, small, and medium enterprises (MSMEs) (Asian Development Bank [ADB], 2023), government can assist through human capacity development in the areas of management, technology, and sustainability. Training and lifelong learning for tourism stakeholders can increase manpower quality, which is beneficial to individuals, organizations, industry, and the economy in general. This allows tourism supply to catch up and meet industry goals (Parsons, 2022) without the need for government to fix the market and continuously justify its policy direction.

However, manpower training can be a double-edged sword because the travel and tourism industry is spatial (Zhao & Liu, 2021), economically and geographically. With labor mobility becoming more liberalized as evidenced by the existence of Mutual Recognition Arrangements (MRAs) in an unregulated industry like tourism (Rivera et al., 2019), increasing the quality of manpower makes them attractive to move to other labor markets due to persistent imbalances and wage differentials (Caliendo et al., 2022). While this situation calls for stringent regulations in managing skills and talents, market failures can exist due to the unintended consequences emerging from regulations (e.g., loopholes in regulation) that may do more harm than good particularly to the vulnerable and marginalized (McLaughlin, 2018). To address this, we reiterate the need for a new mindset in viewing regulation.

Government can pivot toward being more developmental than regulatory (Rivera et al., 2019) by not insisting on fixing the market but rather correct market failures (Thoma, 2016) by creating an environment where new products and new markets can be created. By setting an overall thrust for tourism development on the basis of emerging trends, lessons, and changing preferences, government is able to set the trajectory of where the market should head. This will facilitate the supply of tourism products and services in catching up with the renewed demand for tourism products and services.

The Case of Newport World Resorts: Ecological Civilization

Pursuing sustainability has been at the forefront of tourism development policy in many countries, even prior to the COVID-19 pandemic, more so during and after the pandemic (Rivera et al., 2022; Roxas et al., 2022). This has also been driven by the UN SDG, ASEAN Framework on Sustainable Tourism Development, and the UN Tourism Framework on Measuring Sustainability in Tourism. In the Philippines, superimposing this in the NTDP 2023-2028 in response to rising demand for sustainable products and services, private sector has demonstrated readiness in responding to the call by establishing their own initiatives.

Case in point, Newport World Resorts (NWR) has integrated sustainability program in its entire development, including all hotels in its portfolio (i.e., Manila Marriott Hotel, Sheraton Mania Hotel. Hilton Manila, Hotel Okura Manila, and Holiday Inn Express Manila).[10] Hinging on the existing sustainability programs of each hotel brand, the "I Love Earth" (ILE) program was created that taps into collaborative engagement and personal commitment of every individual (i.e., employee and stakeholders) to put sustainability at the core of professional and personal practice. This is aligned with the concept of ecological civilization (Gu et al., 2020) wherein ecology lies at the heart of politics, economy, culture, and society. Likewise, the ILE is also anchored on a framework with measurable goals under each of the five HEART pillars.[11]

With ecological civilization, social, economic, and governance systems can be designed to promote the overall well-being of the planet and the people, alongside enterprises. Taking the case of the People's Republic of China (PRC), ecological civilization can be understood as their new system of development and governance on the basis of political decision-making anchored on environmental management, ecological restoration, and green development (Xue et al., 2023). Hence, PRC is able to contribute not only to national sustainability but also to global sustainability (Gu et al., 2020).

Through ecological civilization, we approach a more sustainable and just society (i.e. a world that works for everyone). For this to materialize, a paradigm shift is required to realize that the major social and environmental challenges the world is facing are intertwined, and solutions must deviate from the tragedy of the commons (Spiliakos, 2019).

Adapting Community-Based Tourism

Appealing to the ILE's pillar of "Transform through Teaching, Training, Technology," the approach of underscoring participation and effective communication of the commitment to promoting sustainability among employees promotes ownership of the initiative. By providing the necessary training and support to enable them to execute the program; by informing suppliers of the salient details of the sustainability policy and encouraging them to adopt effective sustainability management practices; and by soliciting input from our employees, suppliers, customers, and board members in meeting sustainability goals, there exists not

---

[10]NWR is the first integrated resort in the Philippines; a one-stop, nonstop entertainment and leisure destination that features gaming thrills, world-class entertainment, unique events, and exciting lifestyle options. Official website: https://www.newportworldresorts.com/

[11]For more details, see https://www.newportworldresorts.com/sustainability-policy

only ownership but comanagement of the initiative, which is the guiding principle of community-based tourism (Roxas et al., 2020a; Yanes et al., 2019).

By harnessing the positive externalities of stakeholders having a sense of ownership and comanagement of initiatives, projects, programs, domains, and destinations, government is able to relieve itself from micromanaging everything (Desai, 2021) and instead focus on developing new products and services, finding the right market, stimulating the market, enabling the market to be competitive, protecting the environment and consumers, capacitating manpower, facilitating capital formation, projecting value for accreditation and certification, among others. That is, government can create more value, accelerate sustainability in the market, and foster resilience in the industry.

In the Philippines, not all tourism enterprises are accredited by due to a variety of reasons (Adriano, 2020; Arnaldo, 2017). This became pronounced during the pandemic because "financial aid that will be available to the industry is meant for accredited tourism enterprises" (Tabios, 2020, para. 5). Despite offering free accreditation and threats of revocation, cancellation, or nonissuance of succeeding business permits, accreditation remained challenging. As a consequence of stakeholders having a sense of ownership and comanagement of initiatives, projects, programs, domains, and destinations, accreditation and certification can now be facilitated but with a different approach.

Because government is able to be developmental by going beyond regulation and focus its attention on policy variables that can fuel the travel, tourism, and hospitality industry, the market will find it imperative to self-regulate and commit to accreditation and certification. This is also indicative of the enhanced value of government recognition through accreditation and certification. Enterprises will choose to be accredited and certified because of the value that it offers – standards are kept, product values are aligned, health and safety are ensured, and everyone benefits from a unified value proposition and messaging of a nationwide tourism branding. In the long run, government is able to empower markets to self-regulate that will alleviate the government from conducting excessive regulation on areas that can self-regulate. That is, there is no need for government to fix the market and justify its policy direction. Given the constantly evolving global competitive landscape and the ever increasing alternatives for tourists, consumers, and investors, creating a sustainable and resilient industry with a resonating identity allows countries to be in a strategic position in a highly competitive global landscape (Roll, 2018).

Lessons Learned

We revisited government's regulatory function in tourism development. While government's regulatory nature is a permanent fixture, the nature of regulation can be reimagined. Because of developments in globalization, competition, global travel preferences and behaviors, and economic landscape, we underscored that for government to remain a powerful catalyst of development, it should exceed regulation and pivot toward developmental approaches that can facilitate sustainability and resilience. Hence, it is able to address environmental, social, and governance issues at the same time and give focus on other pressing matters. Thus, government can accelerate sustainability across the tourism value chain.

As a matter of policy, it is important for government to consider two things in any policy formulation: (1) the negative externalities of policies and regulations must be mitigated and (2) asymmetric information must be reduced. With technological development and innovation, free markets have the capacity to fix its own problem without the need for regulation (e.g., negative feedbacks on tourism and hospitality enterprises are not the end because they are given the runway to self-correct). As such, the market can move beyond boundaries as technological development and innovation can push markets to new frontiers, thereby creating new markets. In the words of Josephson (2024), "consumers need free competition, not misguided regulations" (para. 1).

The changing consumer preferences, accompanied by technological innovation, prompt stakeholders to build markets that value sustainability (i.e., if consumers demand sustainable products and tourism enterprises do not go sustainable, then there is no future for tourism).

In realizing this, government now becomes more than a regulator but a change agent. In the Philippines, for example, the DOT reviewed the Tourism Act of 2009 (Republic Act [RA] 9593) for potential amendments and recalibrations since the market was disrupted by the pandemic.[12] It rationalized the function of government toward adapting to the evolving landscape of tourism that was accelerated by the pandemic wherein disruptions allowed the emergence of greater digital tourism, e-commerce adoption, and sustainability in various areas. This gives the industry the much needed thrust to adapt to the new market and not return to the old ways of doing tourism. While private sector can also be a change agent, government can initiate the change through establishing collaboration and coordination with stakeholders toward healthy, sustainable, and resilient market development. While the Boracay Rehabilitation Program of 2018 was seen as the government's demonstration of its regulatory function, it facilitated market adjustments toward sustainability as other destinations self-regulated. This is government's primal role of influencing and shaping the market, instead of fixing the market and justifying policy direction.

New market frontiers require training as technological development and innovation impact the type and quality of products and service tourism offers. Manpower must be ready and adept with changing technology. As such, alongside an updated and upgraded tourism policy, a nationwide approach to upgrading skills is a must.

Therefore, the Philippines can serve as a model in incorporating sustainability into its policy (e.g., PDP 2023-2028, NTDP 2023-2028, review and amendment of RA 9593). Through these policy frameworks, sustainability is not just a plan or a

---

[12]A policy that acknowledges tourism as an "indispensable element of national economy and an industry of national interest and importance." For more information, see https://www.officialgazette.gov.ph/2009/05/12/republic-act-no-9593-s-2009/

strategy but an inherent pillar of any programs, activities, and projects (PAPs) that any stakeholder will implement. That is, the action plans and policies of government are spelled out on how PAPs must be carried out so that sustainability, resilience, and inclusivity are ensued.

These are strategies to achieve DOT's vision "to develop a globally competitive, environmentally sustainable and socially responsible tourism industry that promotes inclusive growth through employment generation and equitable distribution of income thereby contributing to building a foundation for a high trusted society" (DOT, n.d., para. 2). Therefore, through collaboration and cooperation, the travel, tourism, and hospitality industry can be shaped toward sustainability through a combination of sound government policy, reconceptualization of tourism, pivot in marketing approaches and priorities, and the integration of sustainability principles into decision-making processes.

References

Adriano, L. (2020, January 10). Tourism firms reminded of no accreditation, no biz permit policy. *Philippine News Agency*. https://www.pna.gov.ph/articles/1090544

Agafonova, T., & Spektor, L. (2023). Legal regulation of activities in the tourism industry. In A. Beskopylny, M. Shamtsyan, & V. Artiukh (Eds.), *XV international scientific conference "INTERAGROMASH 2022" lecture notes in networks and systems* (Vol. 574, pp. 991–997). Springer. https://doi.org/10.1007/978-3-031-21432-5_104

Allaberganov, A., Preko, A., & Mohammed, I. (2021). Government commitment to tourism and hospitality sector during COVID-19 pandemic. *Tourism Critiques: Practice and Theory*, *2*(2), 153–169. https://www.emerald.com/insight/content/doi/10.1108/TRC-02-2021-0004/

Ammirato, S., Felicetti, A. M., & Della Gala, M. (2014). Tourism destination management: A collaborative approach. In L. M. Camarinha-Matos & H. Afsarmanesh (Eds.), *Collaborative systems for smart networked environments. PRO-VE 2014. IFIP advances in information and communication technology* (Vol. 434, pp. 217–226). Springer. https://doi.org/10.1007/978-3-662-44745-1_21

Andrews, D., Williamson, T., & Aksli, M. (2021, June 25). How will the role of government change after the pandemic? In *Governance for development*. World Bank. https://blogs.worldbank.org/governance/how-will-role-government-change-after-pandemic

Arnaldo, M. S. F. (2017, June 18). Thousands of tourism firms have no DOT accreditation. *Business Mirror*. https://businessmirror.com.ph/2017/06/18/thousands-of-tourism-firms-have-no-dot-accreditation/

Asia-Pacific Economic Cooperation (APEC). (2024, February). *Establishing a safe and sustainable future for travel in the better normal* (Compendium). Asia-Pacific Economic Cooperation. https://www.apec.org/publications/2024/02/establishing-a-safe-and-sustainable-future-for-travel-in-the-better-normal—compendium

Asian Development Bank (ADB). (2023, February 21). Helping tourism MSMEs take the digital path to recovery and resilience. *Southeast Asia Development Solutions Knowledge and Innovation Platform*. https://seads.adb.org/solutions/helping-tourism-msmes-take-digital-path-recovery-and-resilience

Atalla, G., Mills, M., & McQueen, J. (2022, May 13). *Six ways that governments can drive the green transition*. Ernst & Young Global Limited. https://www.ey.com/en_gl/sustainability/green-money-flows

Babii, A., & Nadeem, S. (2021, February 26). Tourism in a post-pandemic world. *IMF News*. International Monetary Fund. https://www.imf.org/en/News/Articles/2021/02/24/na022521-how-to-save-travel-and-tourism-in-a-post-pandemic-world

Bevir, M. (2009). *Key concepts in governance*. Sage. https://uk.sagepub.com/en-gb/eur/key-concepts-in-governance/book230905

Bhuiyan, M. A., Zhang, Q., Xuan, W., Rahman, M. K., & Khare, V. (2023). Does good governance promote sustainable tourism? A systematic review of PESTEL analysis. *SN Business & Economics*, *3*(33). https://doi.org/10.1007/s43546-022-00408-x

Bichler, B. F. (2021). Designing tourism governance: The role of local residents. *Journal of Destination Marketing & Management*, *19*. https://doi.org/10.1016/j.jdmm.2019.100389

Björk, P. (2000). Ecotourism from a conceptual perspective, an extended definition of a unique tourism form. *International Journal of Tourism Research*, *2*(3), 189–202. https://doi.org/10.1002/(SICI)1522-1970(200005/06)2:3%3C189::AID-JTR195%3E3.0.CO;2-T

Bramwell, B., & Lane, B. (Eds.). (2000). *Tourism collaboration and partnerships: Politics, practice and sustainability*. Channel View Publications. https://doi.org/10.21832/9780585354224

Brouder, P., Clavé, S. A., Gill, A., & Ioannides, D. (Eds.). (2017). *Tourism destination evolution*. Routledge Taylor & Francis Group. https://www.routledge.com/Tourism-Destination-Evolution/Brouder-Clave-Gill-Ioannides/p/book/9780367668259

Buhalis, D., & Fletcher, J. (1995). Environmental impacts on tourism destinations: An economic analysis. In H. Coccosis & P. Nijkamp (Eds.), *Sustainable tourism development* (pp. 3–24). Avebury. https://www.semanticscholar.org/paper/Environmental-impacts-on-tourist-destinations%3A-an-Buhalis-Fletcher/0183f32f8ef4ad65113308bf83ec8a5709226d5e

Bushell, R., & Bricker, K. (2016). Tourism in protected areas: Developing meaningful standards. *Tourism and Hospitality Research*, *17*(1), 106–120. https://doi.org/10.1177/1467358416636173

Byrd, E. T. (2007). Stakeholders in sustainable tourism development and their roles: Applying stakeholder theory to sustainable tourism development. *Tourism Review*, *62*(2), 6–13. https://doi.org/10.1108/16605370780000309

Caliendo, M., Künn, S., & Mahlstedt, R. (2022, January). *The intended and unintended effects of promoting labor market mobility* (IZA DP No. 15011). IZA Institute of Labor Economics. https://docs.iza.org/dp15011.pdf

Clifton, D., & Amran, A. (2010). The stakeholder approach: A sustainability perspective. *Journal of Business Ethics*, *98*(1), 121–136. https://doi.org/10.1007/s10551-010-0538-6

Danziger, J. N. (2012). *Understanding the political world: A comparative introduction to political science* (11th ed.). Pearson.

Department of Tourism (DOT). (n.d.). *Mission and vision*. https://beta.tourism.gov.ph/mission-and-vision/

Department of Tourism, Dr. Andrew L. Tan Center for Tourism, and Guide to the Philippines (DOT, ALTCFT, GTTP). (2020). *The Philippine travel survey report: Insights on Filipino travelers' sentiments on the new normal*. Department of Tourism, Asian Institute of Management, Guide to the Philippines. https://guidetothephilippines.ph/ph-travel-survey

Department of Tourism, Dr. Andrew L. Tan Center for Tourism, and Guide to the Philippines (DOT, ALTCFT, GTTP). (2021). *The evolving landscape of domestic travel in the Philippines: A study of travelers and tourism enterprises amidst the pandemic*. Department of Tourism, Asian Institute of Management, Guide to the Philippines.

Desai, N. (2021, July 28). Govt must not micromanage. *Business Standard*. https://www.business-standard.com/article/opinion/govt-must-not-micromanage-121072801860_1.html

Erkuş-Öztürk, H. (2011a). Emerging importance of institutional capacity for the growth of tourism clusters: The case of Antalya. *European Planning Studies*, *19*(10), 1735–1753. https://doi.org/10.1080/09654313.2011.614384

Erkuş-Öztürk, H. (2011b). Modes of tourism governance: A comparison of Amsterdam and Antalya. *Anatolia*, *22*(3), 307–325. https://doi.org/10.1080/13032917.2011.614354

Fabbri, M., & Faure, M. (2018). Toward a "constitution" for behavioral policy-making. *International Review of Economics*, *65*, 241–270. https://doi.org/10.1007/s12232-018-0296-9

Fennell, D. A., & Malloy, D. C. (1999). Measuring the ethical nature of tourism operators. *Annals of Tourism Research*, *26*(4), 928–943. https://doi.org/10.1016/S0160-7383(99)00032-8

Gössling, S., & Schweiggart, N. (2022). Two years of COVID-19 and tourism: What we learned, and what we should have learned. *Journal of Sustainable Tourism*, *30*(4), 915–931. https://doi.org/10.1080/09669582.2022.2029872

Gowreesunkar, V. G., Maingi, S. W., & Korstanje, M. E. (2024). Introduction: The changing landscape of tourist behavior – Navigating from the new normal to the next normal. In S. W. Maingi, V. G. Gowreesunkar, V. G., & M. E. Korstanje (Eds.), *Tourist behaviour and the new normal* (Vol. I, pp. 1–7). Palgrave Macmillan. https://doi.org/10.1007/978-3-031-45848-4_1

Graci, S. (2013). Collaboration and partnership development for sustainable tourism. *Tourism Geographies*, *15*(1), 25–42. https://doi.org/10.1080/14616688.2012.675513

Graham, J., Amos, B., & Plumptre, T. (2003). *Governance principles for protected areas in the 21st century*. Institute on Governance. https://www.files.ethz.ch/isn/122197/pa_governance2.pdf

Gu, Y., Wu, Y., Liu, J., Xu, M., & Zuo, T. (2020). Ecological civilization and government administrative system reform in China. *Resources, Conservation and Recycling*, *155*. https://doi.org/10.1016/j.resconrec.2019.104654

Gutierrez, E. L. M. (2021). Redefining sustainable tourism in COVID-19: A political-economy approach. *DLSU Business & Economics Review*, *31*(2), 1–13. https://www.dlsu.edu.ph/wp-content/uploads/2021/04/1Gutierrez-041121.pdf

Gutierrez, E. L. M., Rivera, J. P. R., Roxas, F. M. Y., & Zamora, M. L. (2020, September/October). Rebooting Philippine tourism from the COVID-19 pandemic. *The World Financial Review*, 28–32. https://worldfinancialreview.com/rebooting-philippine-tourism-from-the-covid-19-pandemic/

Haiying, C., Linsi, H., Peng, L., Xiaohong, Z., & Changjiang, Y. (2018). Relationship of stakeholders in protected areas and tourism ecological compensation: A case study of Sanya coral reef national nature reserve in China. *Journal of Resources and Ecology*, *9*(2), 164–173. https://doi.org/10.5814/j.issn.1674-764x.2018.02.006

Hall, C. M. (2005). The role of government in the management of tourism: The public sector and tourism policies. In L. Pender & R. Sharpley (Eds.), *The management of tourism*. Sage. https://doi.org/10.4135/9781446214961.n15

Jamal, T., & Camargo, B. A. (2014). Sustainable tourism, justice and an ethic of care: Toward the just destination. *Journal of Sustainable Tourism*, *22*(1), 11–30. https://doi.org/10.1080/09669582.2013.786084

Josephson, K. (2024, March 5). Regulating 'junk' fees is not the way to help bank customers. *FEE Stories*. https://fee.org/articles/regulating-junk-fees-is-not-the-way-to-help-bank-customers/?fbclid=IwAR2AGKN0R1drsZ1KC7ASIBJqbqFamDWRsUaGe_CYEsjjf9rf1wBCSk98jfc

Koscak, M., & O'Rourke, T. (2017). A multi-stakeholder approach to tourism development. In J. Ateljevic & S. J. Page (Eds.), *Tourism and entrepreneurship* (pp. 261–279). Routledge. https://parc.ipp.pt./index.php/sensos/article/download/3484/1465/4392

Lee, C. C., Chen, M. P., Wu, W., & Xing, W. (2021). The impacts of ICTs on tourism development: International evidence based on a panel quantile approach. *Information Technology & Tourism*, *23*, 509–547. https://doi.org/10.1007/s40558-021-00215-4

Luo, J. M., & Zhang, H. Q. (2015). Rule and regulation, tourism. In J. Jafari & H. Xiao (Eds.), *Encyclopedia of tourism*. Springer. https://doi.org/10.1007/978-3-319-01669-6_621-1

McLaughlin, P. (2018, April 26). *Regulation's unintended consequences can hurt everyone – The poor most of all*. Mercatus Policy Briefs. Mercatus Center, George Mason University. https://www.mercatus.org/research/policy-briefs/policy-spotlight-regulations-unintended-consequences-can-hurt-everyone-poor

Miller, G., & Twining-Ward, L. (2005). *Monitoring for a sustainable tourism transition: The challenge of developing and using indicators*. CABI Publishing. https://www.cabidigitallibrary.org/doi/book/10.1079/9780851990514.0000

O'Grady, N., & Shaw, D. (2023). Resilience, responsibility and state abandon: The changing role of the government in emergencies. *Political Geography*, *100*. https://doi.org/10.1016/j.polgeo.2022.102796

Organisation for Economic Co-operation and Development (OECD). (2020, September 2). Building resilience to the Covid-19 pandemic: The role of centres of government. *OECD Policy Responses to Coronavirus (COVID-19)*. https://www.oecd.org/coronavirus/policy-responses/building-resilience-to-the-covid-19-pandemic-the-role-of-centres-of-government-883d2961/

Parsons, L. (2022, August 23). *Why is professional development important?* Harvard Division of Continuing Education. https://professional.dce.harvard.edu/blog/why-is-professional-development-important/

Provan, K., & Kenis, P. (2007). Modes of network governance: Structure, management, and effectiveness. *Journal of Public Administration Research and Theory*, *18*, 229–252. https://doi.org/10.1093/jopart%2Fmum015

Rakhmawati, A., Kusumawati, A., Rahardjo, K., & Muhammad, N. (2020). The role of government regulation on sustainable business and its influences on performance of medium-sized enterprises. *Journal of Sustainability Science and Management*, *15*(2), 162–178. https://jssm.umt.edu.my/wp-content/uploads/sites/51/2020/05/13.15.2pdf.pdf

Ray, S. (2020). A pragmatic approach of interaction between technology and tourism-hospitality. In A. Hassan & A. Sharma (Eds.), *The Emerald handbook of ICT in tourism and hospitality* (pp. 19–29). Emerald Publishing Limited. https://doi.org/10.1108/978-1-83982-688-720201002

Rivera, J. P. R., Cudia, C. P., & Tullao, Jr., T. S. (2019). *Assessing the readiness of Filipino MRA-supported professions to participate in the mobility of skilled labor in the ASEAN region: Lessons for APEC economies* (Discussion Paper Series No. 2019-12). Philippine Institute for Development Studies. https://pidswebs.pids.gov.ph/CDN/PUBLICATIONS/pidsdps1912.pdf

Rivera, J. P. R., Gutierrez, E. L. M., & Roxas, F. M. Y. (2022). Re-thinking governance in tourism: Harnessing tourism's post-COVID-19 economic potential. *Journal of Quality Assurance in Hospitality & Tourism*. https://doi.org/10.1080/1528008X.2022.2135164

Robina-Ramirez, R., Sánchez-Oro Sánchez, M., Jiménez-Naranjo, H., & Castro-Serrano, J. (2022). Tourism governance during the COVID-19 pandemic crisis: A proposal for a sustainable model to restore the tourism industry. *Environment, Development and Sustainability*, *24*(5), 6391–6412. https://doi.org/10.1007%2Fs10668-021-01707-3

Rodolfo, M. C. S., Roxas, F. M. Y., Gutierrez, E. L. M., & Rivera, J. P. R. (Eds.). (2022). *People, planet, profit: Principles and practices of sustainable tourism*. De La Salle University Publishing House.

Roll, M. (2018, December). Country branding strategies for nations and companies. https://martinroll.com/resources/articles/branding/country-branding-strategies-for-nations-and-companies/

Roxas, F. M. Y., Rivera, J. P. R., & Gutierrez, E. L. M. (2020a). Framework for creating sustainable tourism using systems thinking. *Current Issues in Tourism*, *23*(3), 280–296. https://doi.org/10.1080/13683500.2018.1534805

Roxas, F. M. Y., Rivera, J. P. R., & Gutierrez, E. L. M. (2020b). Mapping stakeholders' roles in governing sustainable tourism destinations. *Journal of Hospitality and Tourism Management*, *45*, 387–398. https://doi.org/10.1016/j.jhtm.2020.09.005

Roxas, F. M. Y., Rivera, J. P. R., & Gutierrez, E. L. M. (2022). Bootstrapping tourism post-COVID-19: A systems thinking approach. *Tourism and Hospitality Research*, *22*(1), 86–101. https://doi.org/10.1177%2F14673584211038859

Spiliakos, A. (2019, February 6). Tragedy of the commons: What it is and 5 examples. *Harvard Business School Online*. https://online.hbs.edu/blog/post/tragedy-of-the-commons-impact-on-sustainability-issues

Tabios, H. (2020, July 14). DOT accreditation a must for tourism businesses-TCP. *Manila Bulletin*. https://mb.com.ph/2020/07/13/dot-accreditation-a-must-for-tourism-businesses-tcp

Thoma, M. (2016, May 3). When markets aren't perfect, government can help. *CBS News*. https://www cbsnews.com/news/when-markets-arent-perfect-government-can-help/

Tosun, C. (2006). Expected nature of community participation in tourism development. *Tourism Management*, *27*(3), 493–504. https://doi.org/10.1016/j.tourman.2004.12

United Nations World Tourism Organization (UNWTO). (2013). *Sustainable tourism for development guidebook*. United Nations World Tourism Organization. https://www.e-unwto.org/doi/pdf/10.18111/9789284415496

Wan, Y. K. P., Li, X., Lau, V. M.-C., & Dioko, L. (2022). Destination governance in times of crisis and the role of public-private partnerships in tourism recovery from Covid-19: The case of Macao. *Journal of Hospitality and Tourism Management*, *51*, 218–228. https://doi.org/10.1016/j.jhtm.2022.03.012

Whitworth, E. (2023, March 16). Is market self-regulation a thing? Adam Smith thinks so. *Shortform*. https://www.shortform.com/blog/is-market-self-regulation-a-thing/

Xue, B., Han, B., Li, H., Gou, X., Yang, H., Thomas, H., & Stückrad, S. (2023). Understanding ecological civilization in China: From political context to science. *Ambio: A Journal of Environment and Society*, *52*, 1895–1909. https://doi.org/10.1007/s13280-023-01897-2

Yanes, A., Zielinski, S., Cano, M. D., & Kim, S. (2019). Community-based tourism in developing countries: A framework for policy evaluation. *Sustainability*, *11*(9). https://doi.org/10.3390/su11092506

Zhao, M., & Liu, J. (2021). Study on spatial structure characteristics of the tourism and leisure industry. *Sustainability*, *13*(23). https://doi.org/10.3390/su132313117

Epilogue

Maria Cherry Lyn S. Rodolfo

In the postpandemic era, sustainability has been mainstreamed in policies, strategies and practices of enterprises, destinations, and government. The better normal emerged because of paradigm shifts. Value creation is prioritized over volume generation to create sustainable jobs and investments. Customization of products and services to meet the growing demand for sustainability by the visitors has emerged as a core program of destinations. Resources have been calibrated toward products and markets that generate highest economic, social, and environmental returns, when matched with destination readiness and capabilities.

Tourism has become more inclusive across the value chain of primary and support activities, demonstrating its power and potentials to create and magnify opportunities and returns for stakeholders involved in the exchange. No author ever mentioned that value chain upgrading toward sustainable tourism would be easy. It is quite the opposite. Challenges in terms of lack of resources and capacity and presence of information asymmetry and moral hazards have been encountered along the way. MSMEs have become more resilient because of investments in their own innovation and their collaboration with the bigger players and the government. Institutional and governance innovations serve to create an improved system of incentives for stakeholders, small to large groups, private sector, and community groups, to invest their resources and participate in tourism by addressing those challenges.

The footprints of tourism, big or small, positive or negative, are highly visible at the destination level – in the "small" places like Intramuros, Binondo, Samar, and Nueva Ecija that the authors revisited during their journey. These "small" places and initiatives have transformed into resilient "big" and global destinations. The authors have used the power of data – proper collection, analysis, and presentation – to show how the "smallness" of the destination can actually make big impacts on sustainability. Data prove to be critical in the due diligence process.

Indeed, it is a daunting task to ensure that no one is left behind, that Philippine tourism's trajectory is aligned to the Sustainable Development Goals and the country's *Ambisyon 2040* for Filipinos: *Matatag, Maginhawa at Mapanatag na Buhay* (by 2040, Filipinos enjoy a strongly rooted, comfortable, and secure life).[1] But it can be done. Women are strong advocates and champions in environmental and cultural protection and conservation. The informal sector (e.g., Binondo, Manila) is an essential part in the inclusive tourism business as a result of the opportunities, right incentives, and capacity development investments. Property rights system has become instrumental in defining boundaries and incentives for resource allocation, access, and use, especially in protected areas. Sustainable tourism in the Philippines thrives because these property rights are well-defined. Sustainable tourism is the major partner in financing the conservation of our tourism resources for the future generation.

Beyond the better normal described above, sustainable tourism is supported by a productive and resilient pool of talents, able to cope with the challenges of technological advancements and climate change impacts. Tourism networks even though complex are widespread, strong, and effective in reducing the disparities in development between urban and rural destinations and between small and large players. They are instrumental in spreading tourism's benefits, in clustering resources for shared growth, and in making destinations smarter and sustainable. How these elements have evolved and woven to become part of the better normal story is the subject of another journey.

[1]https://2040.neda.gov.ph/about-ambisyon-natin-2040/

Index